DYNAMIC ISSUES IN
CAPITALIST DEVELOPMENT

DYNAMIC FORCES IN CAPITALIST DEVELOPMENT

A Long-Run Comparative View

ANGUS MADDISON

Oxford   New York

OXFORD UNIVERSITY PRESS

Oxford University Press, Walton Street, Oxford OX2 6DP

Oxford New York Toronto
Delhi Bombay Calcutta Madras Karachi
Kuala Lumpur Singapore Hong Kong Tokyo
Nairobi Dar es Salaam Cape Town
Melbourne Auckland Madrid

and associated companies in
Berlin Ibadan

Oxford is a trade mark of Oxford University Press

Published in the United States
by Oxford University Press Inc., New York

British Library Cataloguing in Publication Data
Data available

Library of Congress Cataloging in Publication Data
Maddison, Angus.
Dynamic forces in capitalist development: a long-run comparative
view/Angus Maddison.
p. cm.
Includes bibliographical references and index.
1. Economic history. 2. Capitalism—History. 3. Industrial
productivity—History. 4. Economic policy—History. I. Title.
HC51.M25 1991   338.9—dc20   91–12836
ISBN 0–19–828397–0
ISBN 0–19–828398–9 (Pbk.)

3 5 7 9 10 8 6 4

Printed in Malta

ACKNOWLEDGEMENTS

I am grateful to Moses Abramovitz, Bart van Ark, André Hofman, Dirk Pilat, Gunnar Viby Mogensen, and Eddy Szirmai for comments on many issues raised in this book. Nanno Mulder provided excellent statistical assistance and Dirk Pilat prepared the graphs.

On historical statistics and international comparisons I received helpful advice and comments from Gert den Bakker, Derek Blades, Noel Butlin, Albert Carreras, Edwin R. Dean, Michèle Fleury-Brousse, Jean Gadisseur, Martine Goossens, Roland Granier, Bryan Haig, Akira Hayami, Riitta Hjerppe, Olle Krantz, Hugo Krijnse Locker, John C. Musgrave, Malcolm Urquhart, Vera Zamagni, and Jan Luyten van Zanden.

This book draws on ideas and evidence I have published elsewhere, particularly 'Origins and Impact of the Welfare State, 1883–1983', *Banca Nazionale del Lavoro Quarterly Review*, March 1984, 'Growth and Slowdown in Advanced Capitalist Economies', *Journal of Economic Literature*, June 1987, and *The World Economy in the Twentieth Century*, OECD Development Centre, Paris, 1989.

I profited from presenting some of these ideas at seminars in the European University Institute in Florence, the International Development Centre of Japan, the Australian National University, and at the Universities of Berlin, Bielefeld, Bologna, Helsinki, Geneva, Groningen, Lund, and Venice.

I am grateful to Tineke Tadema, and Elizabeth and Penny Maddison for typing innumerable drafts of the manuscript and tables.

CONTENTS

LIST OF TABLES AND GRAPHS   xi

INTRODUCTION   1

1. INTERPRETING CAPITALIST DEVELOPMENT   5

Distinctive Features and Dimensions of Capitalist
Development   5
The Nature of our Causal Schema   10
Interpreters of the Protocapitalist Epoch   11

    Malthus   11
    Adam Smith   15

Interpreters of the Capitalist Epoch   16

    Ricardo   17
    Marx   18
    Schumpeter   20
    Contemporary Insights into the Development Process   22

    Notes   27

2. CHANGES IN ECONOMIC LEADERSHIP   30

The Dutch Case   30
The British Case   35
The US Case   40
Notes   45

3. LONG-RUN DYNAMIC FORCES IN   48
CAPITALIST DEVELOPMENT

Institutions   52
Natural Resources   56
Demographic Change   60
Growth of Labour Supply   63
Human Capital   63

Physical Capital   65
Technical Progress   66
Structural Change   73
International Trade   74
The Role of Government   76
Colonialism   80
Notes   82

4. FLUCTUATIONS IN THE MOMENTUM OF
   GROWTH   85

Cycle Analysis   85
Long-Wave Analysis   89

   Kondratieff   89
   Kuznets and Abramovitz   96
   Schumpeter   102
   Long-Wave Revivalists   105
   Conclusions on Long-Wave Theories   108

Phases of Growth   111
Notes   125

5. ACCELERATION, SLOW-DOWN, AND
   CONVERGENCE AFTER 1950   128

The Growth Accounts   132
The Quantity and Quality of Labour Input   136

   Unemployment   136
   Labour Hoarding   136
   Hours per Person   137
   Educational Level   138
   Sex-Mix   138
   Augmented Labour Input   139

The Quantity and Quality of Capital Input   139

   Non-Residential Capital   139
   Embodied Technical Progress, the Age Effect, and
      the Quality of Capital   141
   Residential Capital   144

Total Factor Input   145
Forces Supplemental to Factor Input   145

   Impact of Foreign Trade   147
   Impact of Structural Change   148
   Accelerated Diffusion of Technology   152
   Economies of Scale   153
   Impact of the Energy Crisis   153
   Effects of Natural Resource Discovery   155

Extent to which Growth is Explained   159
Extent to which Acceleration and Deceleration are
  Explained   159
Extent to which Convergence is Explained   160
Notes   165

6. THE ROLE OF POLICY IN ECONOMIC
   PERFORMANCE SINCE 1950   167
Characteristics of the Golden Age (1950–1973)   168

   Managed Liberalism in International Transactions   168
   Governmental Promotion of Domestic Demand   169
    Buoyancy
   The Role of Policy and Circumstance in Moderating
    Price Increases   173

The Breakdown of the Golden Age   177

   The Collapse of the Bretton Woods International
    Monetary System   177
   The Erosion of Price Constraints   179
   The Oil Shock   181

Changed Policy Aspirations   182
Adequacy of the Response to the New Challenges   185

   Meeting the OPEC Threat   185
   Deceleration of Inflation   187
   Viability of International Payments Arrangements   187

The Degree of Policy Success since 1973   190
Notes   193

Appendix A   Sources and Methods Used to Measure
              Output Levels and GDP Growth   195
Appendix B   Population   223
Appendix C   Labour Input and Labour Productivity   243
Appendix D   Non-Residential Reproducible Tangible
              Fixed Capital Stock   278
Appendix E   Cost of Living   293
Appendix F   The Volume of Merchandise Exports   308

INDEX   329

TABLES AND GRAPHS

TABLES

1.1 Levels of GDP per Head of Population, 1820–1989   6

1.2 Population, 1500–1989   7

1.3 Comparative Performance of Western Europe (and its
     Offshoots) and China, 1400–1989   8

1.4 Ultimate and Proximate Elements Explaining Per Capita
     GDP Performance   12

1.5 Characteristics of the Advanced Capitalist Countries and
     the Rest of the World, 1900–1989   24

2.1 Employment Structure in the Lead Countries, 1700–1989   32

2.2 Growth of Productivity and Capital Stock in the UK and
     USA, 1700–1989   38

2.3 Ratio of Gross Fixed Non-Residential Investment to GDP
     at Current Market Prices, 1871–1987   41

2.4 Characteristics of Lead Country and Follower Country
     Performance in 1987   43

3.1 Growth Rates of Real GDP per Head of Population,
     1820–1989   49

3.2 Growth Rates of Real GDP, 1820–1989   50

3.3 Phases of Productivity Growth (GDP per Man-Hour),
     1870–1987   51

3.4 Comparative Levels of Productivity, 1870–1987   53

3.5 Total Area Per Head of Population   57

3.6 The Pattern of US Primary Energy Requirements,
     1890–1989   59

3.7 Rates of Population Growth, 1500–1989   62

3.8 Average Years of Formal Educational Experience of the
     Population Aged 15–64 in 1913 and 1989   64

3.9 Gross Non-Residential Fixed Capital Stock per Person   66
     Employed, 1890–1987

3.10 Ratio of Gross Non-Residential Capital Stock to GDP,
     1890–1987   67

3.11 Gross Residential Capital Stock per Head of Population,
     1950–1987   69

3.12   Ratio of Gross Residential Capital Stock to GDP,
        1950–1987   70
3.13   Variations in Productivity Growth in the USA, 1890–1989   71
3.14   Percentage Structure of Employment, 1870–1987   73
3.15   Volume of Exports, 1720–1989   75
3.16   Electorate as Percentage of Persons Aged 20 and Over 77
3.17   Total Government Expenditure as a Percentage of GDP at
        Current Prices, 1913–1987   77
3.18   Structure of Government Expenditure as a Percentage of
        GDP, 1981   79
3.19   Categories of Government Revenue as a Percentage of
        GDP at Current Market Prices, 1987   81
4.1   Amplitude of Recessions in Aggregate Output, 1870–1989   87
4.2   Amplitude of Recessions in Exports, 1870–1989   88
4.3   Kondratieff's Long-Wave Chronology   95
4.4   Amplitude and Duration of Cycles in Industrial
        Production (including Construction), 1870–1913   98
4.5   Schumpeter's Long-Wave Chronology   104
4.6   Mandel's Evidence Scrutinized   107
4.7   Year-to-Year Percentage Change in Aggregate GDP of
        the Sixteen Coutries, 1871–1989   113
4.8   Incidence of Recessions, 1870–1989   115
4.9   Growth Characteristics of Different Phases, 1870–1989   118
4.10   Cyclical Characteristics of Different Phases, 1870–1989   119
4.11   System Characteristics of Different Phases   120
5.1   Growth Acceleration and Slow-down after 1950   129
5.2   Rate of Divergence/Convergence of 15 Follower
        Countries towards Levels of Productivity in the Lead 130
        Country, 1890–1987
5.3   Growth of Labour Potential and Input, 1913–1987   134
5.4   Rate of Growth of Gross Non-Residential Capital Stock,
        1890–1987   140
5.5   Rate of Growth of Net Non-Residential Capital Stock,
        1913–1987   141
5.6   Rate of Growth of Gross Non-Residential Capital Stock
        per Person Employed, 1890–1987   142

5.7 Rate of Growth of Net Non-Residential Capital Stock per Person Employed, 1913–1987 142

5.8 Average Age of Gross Non-Residential Capital Stock, 1890–1987 143

5.9 Rate of Growth of Gross Residential Capital Stock, 1890–1987 145

5.10 Growth of Capital Inputs, and Augmented Joint Factor Input, 1913–1987 146

5.11 Rate of Growth in Volume of Foreign Trade, 1913–1987 148

5.12 Ratio of Commodity Exports plus Imports to GDP at Current Market Prices, 1913–1987 149

5.13 Sectoral Labour Productivity Growth (Gross Value Added per Person Employed), 1913–1987 150

5.14 Level of Output per Person Employed by Sector, 1987 151

5.15 Research and Development Expenditure per Person Employed, 1960–1987 152

5.16 Energy Consumption, 1913–1987 154

5.17 Rate of Growth of Energy Inputs, 1913–1987 154

5.18 Changes in Energy Input Ratios, Energy Economy Relaxation, and Reinforcement, 1913–1987 156

5.19 Identifiable Forces Explaining GDP Growth, 1913–1987 157

5.20 Percentage of Growth which is Explained, 1913–1987 160

5.21 Degree to which GDP Growth, Acceleration, and Deceleration are Explained, 1950–1987 161

5.22 Indentifiable Forces Explaining Labour Productivity Convergence, 1950–1987 162

6.1 Indicators of Macroeconomic Performance, 1870–1989 167

6.2 Indicators of Demand and Price Pressure, 1921–1989 170

6.3 Average Rates of Change in Consumer Price Level, 1870–1989 174

6.4 Movement in Export Prices (in National Currencies), 1950–1989 175

6.5 Key Commodity Prices, 1950–1989 176

6.6 Total International Reserves, 1950–1987 178

6.7 Summary Indicators of Fiscal and Monetary Posture, 1960–1989 183

6.8	Current Balance of Payments as a Percentage of GDP in Current Prices, 1961–1989	186
6.9	The Course of Inflation, 1973–1989	188
6.10	Growth of GDP per Capita, 1973–1981 and 1981–1989	191

APPENDIX TABLES

A.1	1985 Levels of GDP at 1985 US Relative Prices	197
A.2	Gross Domestic Product in 1985 US Relative Prices	198
A.3	Gross Domestic Product NOT Adjusted for Territorial Change	199
A.4	Effect of Change from ICP II to ICP V Bench-Mark on Relative Standing of Individual Countries	200
A.5	Movement in GDP, 1700–1869	206
A.6	Movement in GDP, 1870–1912	208
A.7	Movement in GDP, 1913–1949	210
A.8	Movement in GDP, 1950–1989	216
B.1	Population, 1500–1860	226
B.2	Mid-Year Population, Annual Data, 1870–1913	228
B.3	Mid-Year Population, Annual Data, 1913–1949	232
B.4	Mid-Year Population, Annual Data, 1950–1989	236
B.5	Net Migration, 1870–1987	240
B.6	Vital Statistics, 1820–1987	241
B.7	Mid-Year Population Adjusted to 1989 Boundaries, Bench-Mark Years, 1820–1989	242
C.1	A Comparison of Demographic Structures in 1870 and 1987	244
C.2	Females as a Proportion of the Labour Force, 1910–1987	245
C.3	Labour Force Participation Rates around 1950	246
C.4	Labour Force Participation Rates in 1987	247
C.5	Structure of Employment in 1870, 1950, 1973, and 1987	248
C.6	(a) Unemployment as a Percentage of the Total Labour Force, 1920–1938	260
	(b) Unemployment as a Percentage of the Total Labour Force, 1950–1989	262
C.7	Total Labour Force, 1870–1989	266
C.8	Total Employment, 1870–1989	268

C.9   Annual Hours Worked per Person, 1870–1987   270

C.10   Total Hours Worked per Year, 1870–1989   272

C.11   Productivity 1870–1989: GDP per Man-Hour in 1985 US Relative Prices   274

C.12   Derivation of UK Productivity Levels, 1700–1913   276

C.13   Derivation of Dutch Productivity Levels, 1580–1870   277

D.1   Average Life Expectation of Non-Residential Fixed Capital Assets in Official Estimates   279

D.2   Average Life Expectation of Total Non-Residential Fixed Assets in Standardized Estimates Valued at 1985 US$ 281

D.3   Shares of Machinery and Equipment in Standardized Estimates of Gross Fixed Capital Stock Valued in 1985 US$ 282

D.4   Confrontation of Official Estimates of Total Tangible Non-Residential Fixed Capital Stocks in National Prices and Standardized Estimates   283

D.5   Gross and Net Tangible Non-Residential Fixed Capital Stock, France, 1950–1987   285

D.6   Gross and Net Tangible Non-Residential Fixed Capital Stock, Germany, 1950–1987   286

D.7   Gross and Net Tangible Non-Residential Fixed Capital Stock, Japan, 1890–1987   287

D.8   Gross and Net Tangible Non-Residential Fixed Capital Stock, Netherlands, 1950–1987   289

D.9   Gross and Net Tangible Non-Residential Fixed Capital Stock, UK, 1890–1987   290

D.10   Gross and Net Tangible Non-Residential Fixed Capital Stock, USA, 1890–1989   291

E.1   Consumer Price Indices, 1820–1860   295

E.2   Consumer Price Indices, Annual Data, 1870–1914   296

E.3   Consumer Price Indices, Annual Data, 1914–1950   300

E.4   Consumer Price Indices, Annual Data, 1950–1989   304

F.1   Volume of Exports, 1720–1860   311

F.2   Volume of Exports, Annual Data, 1870–1913   312

F.3   Volume of Exports, Annual Data, 1913–1950   316

F.4   Volume of Exports, Annual Data, 1950–1989   320

F.5   Value of Merchandise Exports f.o.b. at Current Prices and Exchange Rates   324

F.6   Merchandise Exports at 1985 Prices and Exchange Rates   325

F.7   Ratio of Merchandise Exports to GDP at Current
      Market Prices   326

F.8   Ratio of Merchandise Exports to GDP at 1985 Prices   327

GRAPHS

2.1   Locus of Productivity Leadership, 1980–1989   31

3.1   Binary Comparisons of Per Capita GDP Growth,
      1890–1989   55

3.2   Swedish Vital Statistics, 1736–1987   61

3.3   Growth of Non-Residential Gross Capital Stock and GDP
      in the USA and Japan, 1890–1987   68

4.1   A Comparison of Volume Movement of GDP in Different
      Periods, 1870–1989   91

4.2   A Comparison of Volume Movement of Commodity
      Exports in Different Periods, 1870–1989   93

4.3   Sectoral Distribution of Employment at Different Levels
      of Real Per Capita GDP, 1870–1987   109

INTRODUCTION

Since 1820 the advanced capitalist countries have increased their total product seventyfold and now account for half of world GDP. Their real per capita income is now fourteen times what it was in 1820, and six times as high as the average for the rest of the world.

This study analyses the nature of Western progress, using a standardized framework of comparative growth accounts. It tries to identify the causal factors responsible for this unprecedented growth. The explanation is necessarily complex as the range of supply side, demand side, and policy influences has been manifold and advance has not been steady. There was an impressive and very general acceleration of growth in the post-war golden age which ended in 1973. Since then the pace has slowed considerably, though in most of the countries, GDP per capita and labour productivity growth are still above the norms in the thirteen decades from 1820 to 1950.

Over the past century, the USA has had a clear leadership position in GDP per head and labour productivity. Its capital stock per head has been well above that in the fifteen 'follower' countries. Its lead in terms of education, R. & D. and technical innovation has also been clear. In the post-war period, the European countries and Japan have had great success in catching up with US levels of performance. Since 1973, US productivity growth has faltered markedly, for reasons that are difficult to identify, but which strongly suggest that there has been a slowdown in technical progress. Ultimately this may portend a slowdown in future growth for this group of countries as a whole unless new leaders emerge with greater vigour than the USA.

The first chapter presents my interpretation and causal schema of the dynamic forces in capitalist development. It puts them in historical and comparative perspective and includes a commentary on the theories of Smith, Malthus, Ricardo, Marx, Schumpeter, and modern analysts who have made significant contributions to diagnosis of the capitalist process.

Chapter 2 analyses the characteristics of the three successive lead countries since 1600: the Netherlands, the UK, and the USA.

Chapter 3 provides detailed quantitative evidence on the major features which help to explain or are associated with the long-term dynamic character of the capitalist experience.

Chapter 4 presents my interpretation of the four major phases of development since 1820, together with a critique of other authors such as Kondratieff or Schumpeter who have attempted to identify regular long swings in the tempo of development. In my explanation, historical accidents — 'system shocks' — are important and the role of policy error and success is also emphasized.

Chapter 5 uses detailed growth accounting techniques for six of the biggest countries to analyse in detail the reasons for growth acceleration in the post-war golden age (1950–73), and for slow-down after 1973. It also analyses the process of convergence by the follower countries on the USA since 1950 — which contrasts sharply with the divergence of 1870–1950.

Chapter 6 analyses the impact of policy on economic performance, and the ways in which policy has been modified by the new challenges which have arisen since 1973.

The present analysis is intended to illuminate current economic issues in advanced capitalist countries, to contribute to their collective economic history, and to augment the evidence for the study of economic growth and its dynamics.

The Appendices present the quantitative evidence for my own analysis. The sources and procedures are described in some detail so that they may also serve as a working tool for those who wish to develop alternative measures or periodization.

This book is a successor volume to earlier studies which I published in 1964 and 1982.[1] The first of these was mainly concerned to explain the great acceleration of growth performance in the post-war period, and to provide a background perspective of the macroeconomic performance of twelve advanced OECD countries from 1870. The second took a much longer view of capitalist perfrmance, with some quantification of the experience of sixteen countries from 1820. It concentrated on major changes (phases) in the momentum of growth. It tried to determine why it had slowed down so generally after 1973 and to ascertain whether the slow-down constituted a major new phase of capitalist performance. Subsequent evidence has confirmed that the slow-down was indeed a new phase of development, distinctly different in character from the post-war golden age, which poses different tasks and options for government policy.

This study contains a further improvement, updating, and standardization of the quantitative data base for analysing growth performance in the long run. It contains a much more comparable assessment of growth and levels of the capital stock (Appendix D) and a more sophisticated analysis of total factor productivity (Chapter 5). It also puts more emphasis on the broader long-run forces making for change over the capitalist period as a whole (Chapter 3).

NOTE ON INTRODUCTION

1. See *Economic Growth in the West*, Allen & Unwin, London, 1964, and *Phases of Capitalist Development*, Oxford University Press, 1982.

1

INTERPRETING CAPITALIST DEVELOPMENT

DISTINCTIVE FEATURES AND DIMENSIONS OF CAPITALIST DEVELOPMENT

This study is concerned with the performance of advanced capitalist countries. The adjective 'capitalist' is appropriate for two reasons. In the first place, property is predominantly in private hands in these countries, and allocation of goods, services, and factors of production (land, labour, and capital) is made mainly through market mechanisms — with capitalists responding to profit signals, workers to wage incentives, and consumers to prices. In the second place, these economies are highly capitalized — their stocks of physical capital, education, and knowledge are large relative to their income flow and huge compared with pre-capitalist societies; the growth of these capital stocks is a major causal element explaining the distinctive character of their performance.

The most striking characteristic of capitalist performance has been a sustained upward thrust in productivity and real income per head, which was achieved by a combination of innovation and accumulation. In this respect, capitalism is very different from earlier social orders whose property and other social institutions were geared to preserve equilibrium and were less able to afford the risks of change.[1]

The 'advanced' capitalist group consists mainly of West European economies and their offshoots (Australia, Canada, and the USA). By world standards their levels of productivity and real income are closely clustered, and their growth has traced a distinctive orbit. The only non-European country to have entered their orbit is Japan. In the contemporary world there are many capitalist economies, but what distinguishes the 'advanced' countries is both their high level of real income, and their much lengthier experience of rising productivity.

TABLE 1.1. *Levels of GDP per Head of Population, 1820–1989 ($ at 1985 US prices)*

	1820	1870	1913	1950	1973	1989	Coefficient of multiplication, 1820–1989
Australia	1,242	3,123	4,523	5,931	10,331	13,584	11
Austria	1,041	1,433	2,667	2,852	8,644	12,585	12
Belgium	1,024	2,087	3,266	4,228	9,416	12,876	13
Canada	n.a.	1,347	3,560	6,113	11,866	17,576	n.a.
Denmark	988	1,555	3,037	5,224	10,527	13,514	14
Finland	639	933	1,727	3,480	9,072	13,934	22
France	1,052	1,571	2,734	4,149	10,323	13,837	13
Germany	937	1,300	2,606	3,339	10,110	13,989	15
Italy	(960)	1,210	2,087	2,819	8,568	12,955	13

Japan	(588)	(618)	1,114	1,563	9,237	15,101	26
Netherlands	1,307	2,064	3,178	4,706	10,267	12,737	10
Norway	(856)	1,190	2,079	4,541	9,346	16,500	19
Sweden	947	1,316	2,450	5,331	11,292	14,912	16
Switzerland	n.a.	1,848	3,086	6,556	13,167	15,406	n.a.
UK	1,405	2,610	4,024	5,651	10,063	13,468	8
USA	1,048	2,247	4,854	8,611	14,103	18,317	17
Arithmetic average	1,002	1,653	2,937	4,693	10,396	14,456	14

Sources: Appendix A and B. Figures are in US dollars at 1985 prices, adjusted for differences in the purchasing power of currencies. They refer to GDP per capita within the geographic boundaries of the years cited. Per capita consumption was probably around 85 per cent of GDP per capita in 1820, and averaged 58 per cent in 1989. Figures in brackets are rough estimates made by extrapolation or inference rather than hard evidence. For Italy, GDP per capita was assumed to move at the same rate from 1820 to 1861 as it did from 1861 to 1890; for Japan, the nature of the estimates for 1820 and 1870 is explained in Appendix A. For Norway it was assumed that the 1820–70 movement was the same proportionately as in Sweden.

Since 1820[2] the total product of the advanced capitalist group has increased seventyfold, population nearly fivefold, and per capita product fourteenfold (see Tables 1.1 and 1.2). Annual working hours have been cut in half and life expectation has doubled. These 17 decades constitute the capitalist epoch, for the pace of advance in peacetime has virtually always been a huge multiple of that in earlier centuries.

The available evidence strongly suggests that in 1820 our group already had higher income levels than the rest of the world, because they had experienced a slow upward crawl in agricultural productivity, in urbanization, and in living standards over several preceding centuries.[3] Per capita income growth was a tiny fraction of that in the capitalist epoch, but this early experience should not be neglected in interpreting subsequent performance, because it was more or less unique to these countries and involved institutional change which had significance for later developments — the creation of a legal framework and accounting conventions for capitalist enterprise, the emergence of international trade networks, the creation of nation states, and the emergence of modern science. In view of this, it seems appropriate to characterize these earlier four centuries of Western experience as 'protocapitalist'.

Whilst there is enough firm evidence to quantify the macroeconomic performance of individual 'advanced' countries back to 1820 (see Table 1.1), it is much more difficult to do this for earlier centuries. For other countries (see Table 1.5), it is as yet difficult to push the record beyond 1900. Nevertheless, there is some evidence from which one can make rough inferences about the situation in earlier times in Europe and elsewhere, and it is useful to put these inferences in quantitative form. This is done in Table 1.3, which gives a rough idea of the performance of Western Europe and its offshoots over the long haul, and contrasts it with Chinese experience.[4]

We know from the work of Needham[5] that Chinese science and technology were superior to those of Europe in the Middle Ages, and it is likely that Chinese levels of productivity and income were also higher than those of Europe at that time. Perkins[6] has produced evidence to back the view that Chinese productivity and income were stagnant in Europe's protocapitalist epoch and for most of its capitalist epoch, even though China's advancing agrarian economy was able to sustain a large expansion in popula-

TABLE 1.2. *Population, 1500–1989*

	1500 ('000)	1700 ('000)	1820 ('000)	1989 ('000)	Coefficient of multiplication, 1820–1989
Australia	0	0	33	16,807	509
Austria	1,420	2,100	3,189	7,618	2
Belgium	1,400	2,000	3,397	9,938	3
Canada	0	15	640	26,248	41
Denmark	600	700	1,097	5,132	5
Finland	300	400	1,169	4,962	4
France	16,400	21,120	30,698	56,160	2
Germany	12,000	15,000	24,905	61,990	2
Italy	10,000	13,200	19,000	57,525	3
Japan	12,000	30,000	31,000	123,116	4
Netherlands	950	1,900	2,355	14,850	6
Norway	300	500	970	4,228	4
Sweden	550	1,260	2,574	8,439	3
Switzerland	650	1,200	1,829	6,723	4
UK	4,400	8,400	21,240	57,202	3
USA	0	251	9,618	248,777	26
Total	60,970	98,046	152,714	709,715	5

Source: Appendix B; 1500–1820 figures for Australia, Canada, and USA exclude indigenous populations.

TABLE 1.3. *Comparative Performance of Western Europe (and its Offshoots) and China, 1400–1989)*

Year	GDP per capita (dollars at 1985 prices)		Population (million)	
	Western Europe and offshoots	China	Western Europe and offshoots	China
1400	430	500	43	74
1820	1,034	500	122	342
1950	4,902	454	412	547
1989	14,413	2,361	587	1,120

Sources: Western Europe and offshoots (i.e. Australia, Canada, and USA): GDP per capita, 1820–89, from Table 1.1; 1400–1820 growth assumed to be around 0.2 per cent a year in the light of S. Kuznets, *Population, Capital, and Growth*, Heinemann, London, 1974, pp. 139 and 167; population from Table B.1 and from B. T. Urlanis, *Rost Naselenie v Evrope*, Ogiz, Moscow, 1941, p. 414. China: GDP per capita, 1900–89, from Table 1.5; 1400–1900 presumed to be stable on the basis of evidence presented in D. H. Perkins, *Agricultural Development in China, 1368–1968*, Aldine, Chicago, 1969; population, 1400–1820, from Perkins, p. 216; 1950–89 from OECD Development Centre sources.

tion. It seems likely that Western Europe edged ahead of Chinese levels of performance somewhere around 1500 and thereafter took over the mantle of world leadership in technology and economic performance.

THE NATURE OF OUR CAUSAL SCHEMA

In assessing the nature of capitalist performance, one can conduct the causal analysis at two levels — 'ultimate' and 'proximate'.[7] The investigation of ultimate causality involves consideration of institutions, ideologies, pressures of socio-economic interest groups, historical accidents, and economic policy at the national level. It also involves consideration of the international economic 'order', exogenous ideologies, and pressures or shocks from friendly or unfriendly neighbours. These 'ultimate' features are all part of the traditional domain of historians. They are virtually impossible to quantify and thus there will always be legitimate scope for disagreement on what is important. 'Proximate' areas of causality are those where measures and models have been developed by economists and statisticians. Here the relative importance of different influences can be more readily assessed. At this level, one can derive significant insight from comparative macroeconomic growth accounts which try to 'explain' growth of output, output per head, or productivity by measuring inputs of labour and capital, availability of natural resources, influences affecting the efficiency with which resources are combined, and benefits derived from transactions with foreign countries. The most difficult problem at this 'proximate' level of explanation is analysis of the role of technical progress, which interacts in a myriad ways with other items included in the growth accounts.

Table 1.4 presents a schema of these two levels of causality and their interaction. I have tried to give due consideration to both of them but the major contribution of this study is to quantify proximate causality.

INTERPRETERS OF THE PROTOCAPITALIST EPOCH

There are two major economists who wrote at the end of the protocapitalist period, and whose views have had very wide influence. Adam Smith (1723–90) published *The Wealth of Nations* in 1776, and Thomas Malthus (1766–1834) his *Essay on the Principle of Population* in 1798. The contrast between these two books could not be more stark, both in content and style. Smith's is a work of scholarly reflection, insight, and massive erudition. Malthus's is a hastily written pamphlet. Smith was optimistic and Malthus very gloomy about economic performance and potential.

Malthus

Although Malthus wrote later than Smith, it is useful to consider him first, because his message was much more primitive. Malthus had a growth schema which considered only two factors of

TABLE 1.4. *Ultimate and Proximate Elements Explaining Per Capita GDP Performance*

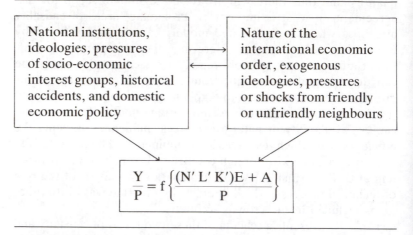

Y = gross domestic product.
P = population.
N′ = natural resources augmented by technical progress.
L′ = human capital, i.e. labour input augmented by investment in education and training.
K′ = stocks of physical capital augmented by technical progress.
E = efficiency of resource allocation.
A = net flow of goods, services, production factors, and technology from abroad.

production — natural resources, and labour — without allowance for technical progress and capital formation. He portrayed the general situation of humanity as one where population pressure put such strains on the ability of natural resources to produce subsistence that equilibrium was attained only by various catastrophes — such as famine, disease, and wars — which brought premature death on a large scale and which he described as 'positive' checks. Later he advocated introduction of 'preventive' checks, such as sexual abstinence, as the only way to avoid such calamities.

Malthus's influence has been strong and persistent, in part because of the forceful rhetoric in which he first couched his simple argument. Thus, while no one would consider his theory as valid for the capitalist epoch, many respectable historians consider it applicable to earlier periods. Here is a sample of his style of argument:

The power of population is so superior to the power in the earth to produce subsistence for man, that premature death must in some shape or other visit the human race. The vices of mankind are active and able ministers of depopulation. They are the precursors in the great army of destruction; and often finish the dreadful work themselves. But should they fail in this war of extermination, sickly seasons, epidemics, pestilence, and plague advance in terrific array, and sweep off their thousands and ten thousands. Should success be still incomplete, gigantic inevitable famine stalks in the rear, and with one mighty blow levels the population with the food of the world.[8]

In fact, the situation in the protocapitalist epoch was not as Malthus suggested, even though most economies were operating under a low-income ceiling because of slow technological progress. The population was indeed subject to mini-famines when bad weather occurred, but not to endemic food shortages lasting over periods of many decades or longer.

There are several reasons for disagreeing with Malthus and his disciples,[9] which are worth stating in view of the persistent popularity of his ideas.

1. European fertility was not at a biological maximum, but already reflected the operation of preventive checks on a significant scale. Unlike Asian countries, Europeans generally lived in conjugal rather than extended families, and in order to sustain living standards marriage did not take place at puberty but in the mid-twenties. Sexual restraint before marriage was enforced with some success by a priesthood which set an example of celibacy, so that a significant fraction of the population was childless. These habits changed temporarily when Europeans emigrated to countries with abundant land, but their existence in protocapitalist Europe has been firmly established by recent French and British demographic research,[10] and one can see in Table 1.3 that the pace of demographic advance in Western Europe from 1400 to 1820 was appreciably slower than in China.

2. 'Average' living standards were well above subsistence. In all countries there was a substantial hierarchy of rulers, upper, and middle classes. The size of this group varied between countries for institutional and political reasons. For England in 1688 we have

Gregory King's estimates which show average income per head of almost £8, but the poorest quarter of the population (cottagers and paupers) survived on a consumption level only 28 per cent of this.[11]

3. There was still a margin of unused land, and it is clear from existing demographic studies that there was migration within Europe. There were much bigger reserves in America, Australia, Siberia, and Africa, which were to offer possibilities for international migration at a later period.

4. The intensity with which land was cultivated could be expanded considerably in most cases by greater per capita labour inputs.[12] In the Middle Ages there were very long off-seasons in which little work was done, and a great deal of land lay fallow. Later generations worked harder, reduced the fallow area, and increased land productivity by pushing agricultural practice closer to that in horticulture.

5. Some of the major demographic set-backs cited by Malthusian pundits for this period (e.g. in the seventeenth century) were not due demonstrably to pressure of population on land but to different causes, such as disease or war. A fundamental weakness of the Malthusian argument is that its central thesis is based on the land–labour dichotomy, but death from disease and war are often used as evidence as if these catastrophes were ultimately due to food shortage.

6. Finally, the protocapitalist economy was not one of complete technological stagnation. Medieval innovations had included windmills, horseshoes, horse harness, heavy ploughs, the haystack, the scythe, marling, fertilization, and the three-field rotation system. These innovations spread rather gradually, but they were certainly helping to increase agricultural output in northern Europe.[13] Innovation was much slower in this epoch than it is now because the main locus of production was in agriculture, where innovation was too risky for most of the participants and often inhibited by tenure institutions. In urban handicrafts, guild restrictions also limited possibilities for change. Entry to skilled occupations was carefully controlled and technical knowledge was regarded as a mystery not to be shared with those outside the recognized confraternity. But the literate group of the population was no longer confined to a priesthood training to conform with tradition rather than to innovate. After the introduction of print-

ing around 1500, diffusion of knowledge was speeded up, and written communication took place in the vernacular rather than in Latin.

Adam Smith

The driving forces of the protocapitalist epoch were brilliantly captured by the explanatory schema of Adam Smith. He emphasized the role of capital deepening in economic growth, the opportunities for economies of scale and specialization, the scope for profitable international trade, and the role that policy could play in accelerating growth. He greatly broadened the significance of the historical approach for growth analysis by using it comparatively. He arrayed countries in an order that corresponds basically with the modern idea of real income per head and built an analytical schema which gave a rough explanation of why they were thus located. There were some drawbacks in his approach; e.g. he did not really distinguish between the benefits accruing from technical progress and economies of scale, he overstressed natural harmony of interest between nations, and he largely ignored the plunder element in the success of the lead country of the epoch.[14]

In fact, it is not surprising that Smith gave much less stress to technical progress than to economies of scale. He was contemplating a range of national economic performance which varied from perhaps $500 average per capita income to $1,500 using our unit of account. His ordering was roughly as follows (excluding what he calls 'naked savages' in pre-agrarian societies):

Netherlands
England
France
British North American colonies
Scotland
Spain
Spanish colonies in America
China
Bengal (depressed by the East India Company's plundering).

Smith was more concerned with policy action that would help push a country nearer to the high-income frontier (located in the Netherlands) by using existing best practice technology than he was with the possibilities for further progress. He gave greater stress to the opportunities arising from removing economic back-wardness than to those from new technology.

Hence he treats capital mainly as a stock which can be increased in per capita terms to make it possible to use more complex methods of production rather than new techniques. In fact, in his definition, capital consisted of a fund to meet the cost of workers' subsistence, as well as provision of tools and equipment. The former element of circulating capital was bigger than the element of fixed capital.[15] It is interesting that his most famous example of the potential gain from exploiting more complex processes relates to the pin-making industry, which was by no means a new one. He was not of course unaware of technical change, but seemed to regard it as a matter of improvement engineering rather than creation of new products or processes.

In discussing policy, Smith stressed the natural harmony of interests of all parties from allocation of resources in free markets, and like his French contemporaries, the Physiocrats, he advocated the case for *laissez-faire* policies of non-intervention. In this he was a very successful advocate because his argument gradually won over British official opinion, and British influence diffused his policy message world-wide.

It should be added that Smith had in mind a rough dichotomy between advanced countries like England and France, where policy changes were regarded as a completely effective way of moving from where they were on the income scale to the frontier position of the Dutch, and the situation in China and India. In the latter group, there were bigger institutional constraints on the adoption of sensible policies, but Smith also had in mind a more moderate view of the Malthus position; i.e. they were, because of more ancient settlement, in a situation where there was greater pressure on natural resources, and less scope for saving. Similarly, he considered North America to be in a different situation, where total output could grow much faster because of the existence of empty land, but where per capita growth and levels would not necessarily be better than in the advanced nations of Europe.

INTERPRETERS OF THE CAPITALIST EPOCH

In order to clarify the present characterization of the driving forces in capitalist development, it may be helpful to comment on the approach adopted by three major economists who advanced their own schema of the capitalist production process: Ricardo (1772–1823), Marx (1818–83), and Schumpeter (1883–1950).

Marx and Schumpeter both had an extremely ambitious approach to capitalist development, involving a socio-political theory, historical and comparative perspective, and vast erudition in the history of economic thought.[16] Ricardo, by contrast, had a minimum of formal education, a narrower perspective, and a greater taste for abstract ideas. Nevertheless, he applied his luminous mind to production of a rigorous schema, which has had an enormous influence on subsequent analysis. He was also more concerned with pragmatic policy questions than Marx and Schumpeter.

Ricardo

Ricardo's schema of capitalist development was adumbrated in 1815, on the eve of the capitalist epoch. He clearly recognized the augmentation in productive power that machinery could bring and the perspectives it offered for substantial economic growth in the non-agricultural sector. However, being strongly influenced by Malthus, he judged that productivity growth was likely to be much slower in agriculture than in industry, because the supply of land was fixed, population was growing, and the increased demand for food would lead to use of less fertile land. As population grew the relative price of food would therefore rise, and this would impact unfavourably on industrial costs. Ricardo assumed that wages tend to be at subsistence level, and thus, when food prices rise, wages must also rise if workers are to survive. As wages rose, industrial profits would be squeezed, and this fall in profits would eventually bring economic expansion to a halt. The obverse of the profit squeeze was the rise in the share of landlords' rents. Thus there was a clash of interests between the new class of industrialists and the landlords. As a temporary relief for this dilemma, Ricardo advocated reduction of duties on imported food, which would keep wages down and postpone the profit squeeze.

Ricardo's argument is advanced in abstract arithmetic terms without reference to history or institutions.[17] Thus it is difficult to categorize his judgement of capitalist potential very clearly. It was certainly more optimistic than that of the young Malthus, and considerably less dynamic than that of Marx and Schumpeter, neither of whom had a two-sector model with scarcity of natural resources as a drag. Some of Ricardo's followers (e.g. J. S. Mill) assumed that his schema implied the advent of economic stagnation within the foreseeable future, and that it would be a stagnation at income levels (for workers) rather near to subsistence. This turned out to be a poor judgement on subsequent capitalist performance, though it can hardly be attributed to Ricardo himself. The other impact of Ricardo has been longer-term. He stimulated the use of two-sector models and theories that expect growth to come to a halt because of natural limits; e.g. Jevons's theory about coal shortage, the Club of Rome's concern with natural limits and ecology, and pessimism about oil shortages. Ricardian thinking about the industrial sector as the more or less exclusive locus of rapid technical change has also had a tremendous influence on subsequent thought, e.g. the notion of an 'industrial' rather than a scientific-technical revolution, the intense concern of many economists with terms of trade between agriculture and industry, or with the 'dangers' of de-industrialization.

Marx

There are many contradictions and paradoxes in the analysis of Marx and Schumpeter, and the following is restricted to what Schumpeter called their 'vision', unencumbered by detail. Marx recognized, more clearly than most of his contemporaries, the enormous productive power of capitalism as compared with that of preceding epochs. He scorned Malthus, and rejected Ricardo's pessimism about the drag to progress associated with the pressure of population on resources. He stressed the enormous growth of productive power represented by the transition from manufacture to machinofacture, and the importance of accelerated accumulation of fixed capital as the mainspring of economic progress. He expected a continued expansion of trade and concentration of production into bigger units, both of which would provide continuing economies of scale. He clearly considered that European plunder and monopolistic trading exploitation of the rest of the world had been a necessary feature of protocapitalism, but he did not make any substantial claim (as some subsequent Marxists have done) that it was a necessary feature of the capitalist epoch.

Marx speculated on the possibility that the momentum of capitalist productive performance might weaken because maintenance of productivity growth might be increasingly inhibited by the difficulty of sustaining technical progress. Sustained progress would require an increased ratio of fixed capital to output, and this might reduce the rate of profit in the long run. However, he put forward several reasons for thinking that this possibility might be offset by countervailing forces. Hence, though Marx expected capitalism's ultimate collapse in favour of socialism, his break-down hypothesis is basically socio-political rather than economic. He expected increasing polarization of the interests of workers and capitalists, and the breakdown was expected as a result of the victory of the worker's interest, which would then abolish private property as a means of production. However, Marx did not present socialism as a stationary state, so under socialism, as with capitalism, Marx presumably would have expected the main-springs of growth to be technical progress and capital accumu-lation. The main difference between capitalism and socialism would be a more equal income distribution, the elimination of unemployment, and a termination of business cycles.

Marx considered a 'reserve army' of unemployed to be a major prerequisite of capitalist economies. It was necessary to keep wages low relative to profits. He did not have a subsistence theory of wages, but a bargaining theory. When demand was high, as in business booms, the bargaining position of labour would improve and profits would be squeezed. In a depression the bargaining position of labour would weaken and the profit outlook would improve. He considered these oscillations in the labour market situation to be the major cause of cyclical fluctuations under capitalism, and he took a ten-year cycle to be typical.[18]

As Marx was not interested in the survival of the capitalist system, he was not really concerned with economic policy, except in so far as the labour movement was involved. There, his argu-ment was concentrated on measures to limit the length of the working day, and to strengthen trade union bargaining power. His analysis was also largely confined to the situation in the leading capitalist country of his day — the UK — and he did not consider the policy problems of other Western countries in catching up with the lead country (as Friedrich List did). In so far as Marx was concerned with other countries, it was mainly with poor countries which were victims of Western imperialism in the merchant capi-talist era.

Schumpeter

Schumpeter gave greater stress to the role of technical progress and less to the role of capital accumulation than Marx did. He rejected completely the Malthus–Ricardo type of constraints arising from pressure of population on fixed natural resources.[19] He also rejected the view that there was any necessary element of imperialist exploitation in capitalist development.[20]

Schumpeter made a sharp distinction between the way an economy would operate as a 'circular flow' if technology were static, and the way it operates in the real world of 'economic development' where 'technique and productive organization' are changing. In a capitalist economy, 'economic life changes its own data by fits and starts'; the system 'so displaces its equilibrium point that the new one cannot be reached from the old one by infinitesimal steps. Add successively as many mail coaches as you please, you will never get a railway thereby.'

Schumpeter did not view the capital stock as the incarnation of technical progress, but stressed the central role of the entrepreneur: 'Capital is nothing but the lever by which the entrepreneur subjects to his control the concrete goods which he needs, nothing but a means of diverting the factors of production to new uses, or of dictating a new direction to production.' He distinguishes sharply between the entrepreneurial role of innovation and that of owning or managing assets. Only the entrepreneur creates profits as distinguished from 'interest', which is the return on ownership. Interest comes in a steady stream, but profits are 'transitory and ever-changing' because the entrepreneur can capture the benefits of innovation only temporarily. Once the viability of the innovation is demonstrated, it will be copied by imitators. It will cease to be an innovation and, having lost its freshness, will drop back into the domain of the circular flow.[21]

Schumpeter regarded innovation as 'difficult and only accessible to people with certain qualities'; 'only a few people have these qualities of leadership'. Hence innovation comes in jerks or 'swarms', discontinuously in time. The economy thus progresses through a series of cycles. One round of innovations gathers momentum as the innovator attracts imitators; then there is stagnation, which is eventually broken by some new entrepreneur.

Thus the entrepreneur is the hero of economic development, and his heroism is all the more legitimate because in each wave new men emerge, as 'the function of the entrepreneur itself cannot be inherited' (p. 79).

Schumpeter described the nature of economic development as the 'carrying out of new combinations', which he defined rather widely as follows (in fact, only the first two of these represents what is conventionally included in the notion of technical progress):

1. introduction of new goods;
2. introduction of new methods of production;
3. opening a new market;
4. conquest of a new supply of raw materials;
5. new organization of an industry.

Schumpeter's provocative approach was a major break with the academic tradition in economics, which had ignored Marx and not taken much interest in growth problems for several decades. He put technical change at the centre stage of capitalist development, and in his discussion of the temporary character of innovation profit, brought out clearly the non-appropriability of knowledge which is the major reason it is so difficult to capture its role in a production function. One aspect of his approach that is difficult to accept is the notion that entrepreneurship is so scarce a factor of production. His own later argument, that innovation can be institutionalized in large firms, itself contradicts the proposition. If the entrepreneur is disenthroned in Schumpeter's schema, then we must fall back on capital as the vehicle for technical change.

Like Marx, Schumpeter was not interested in policy to promote growth in the way in which Adam Smith was, nor did he discuss problems of relative backwardness within the process of capitalist development. Because Smith was so heavily orientated to policy, his unit of analysis was the performance of particular nations. Marx and Schumpeter in their main theoretical work argue in more general terms, so their reference unit is not so clearly national, but in fact they were analysing the capitalist process in the lead country. One reason for the absence of policy discussion is that both Marx and Schumpeter expected the capitalist system to collapse for different reasons. However, it is a little odd that Schumpeter did not discuss patents, R. & D., and invention, which must move one step ahead of the entrepreneurial act. Perhaps he thought that innovation normally occurs well within the frontier of potentially exploitable knowledge.

Contemporary Insights into the Development Process

Apart from Marx and Schumpeter, the literature on economic growth for most of the nineteenth and early twentieth centuries was rather thin. One of the problems of earlier analysts of capitalist development is that they had to work without the benefit of the modern statistical revolution, which owes so much to the intellectual efforts of Simon Kuznets, who developed the analytic framework of national accounts, and encouraged scholars in other countries to produce historical estimates of the major magnitudes. We are now, therefore, much better placed to see when the critical changes in the magnitude of economic growth took place than were earlier writers, using partial indicators such as industrial production or prices, or simply relying on imaginative hypothesis or metaphor. Thanks to pioneers like Colin Clark and Simon Kuznets, we now have an adequate conceptual basis for measuring aggregate economic activity in a national accounting framework. There are official GDP estimates for all our countries since 1950, and reasonably authoritative historical estimates back into the nineteenth century for many of them. Though there is obviously still substantial scope for improvement, the international comparability of these estimates has been enhanced not only by the adoption of common definitions but by extensive empirical work to facilitate comparison of levels of performance by adjustment for differences in the purchasing power of currencies. Here we owe a great deal to the work of Milton Gilbert and Irving Kravis.

There has been a resurgence of interest in economic growth and development in the post-war period, in relation to the problems both of advanced capitalist countries and of the poorer 'developing' countries.

The literature on the advanced countries has been largely technocratic, concerned with models, production functions and proximate causality, without the socio-historic sweep of the Smith–Marx–Schumpeter tradition. Within this literature there have been two important new ideas which have added to the possibility of analysing capitalist development. These are the notions of technical progress being 'embodied' in the capital stock, and of education as a form of 'human capital' embodied in the labour force.

The first of these ideas was presented in its most elaborate form by Salter.[22] He takes capital to be the major vehicle of economic growth because it embodies technical progress. This view leads him to define the capital stock as an accumulation of successive 'vintages' of capital goods, which augment the productive power of investment year by year (whether it be for replacement, widening, or deepening) because of the progress of technique. He makes a distinction between 'best practice' productivity and average productivity, which is extremely helpful in identifying the nature of technological leadership, the reasons why other countries lag behind the productivity leader, and why follower countries can achieve faster growth than the leader. Salter makes a sharp distinction between the contribution to growth of economies of scale and those of technical progress, the former being much less important than the latter.

Another significant development in this literature was the dramatization of the possible significance of education in economic growth with the introduction by Schultz of the concept of 'human capital'.[23] This idea had been adumbrated by Adam Smith, but neglected by Ricardo, Marx, and Schumpeter, who tended to treat all labour as homogeneous. However, the specific identification of the role of education in economic performance is very difficult; and some of the early enthusiasm of human capital pundits who explained wage differentials largely in terms of education, and sought to use the theory to give direct guidelines for educational policy, has met various kinds of scepticism and challenge from authors who think that differences in intelligence, social origin, luck, or credentials have a bigger influence on earnings than has the contribution of education to productivity.[24]

Denison is the most ambitious and successful of the modern analysts who have used production functions to throw light on the relative importance of factors contributing to growth.[25] He does this by giving weights to the items that figure in our Table 1.4, which he derives from the share that each factor has in national product as measured in the national accounts. For each factor, e.g. land, labour, capital, he uses indicators similar to those in our Appendices, except that he disaggregates more. He adjusts labour input for differences in age, sex, and education (*à la* Schultz), but he does not adjust capital stock (*à la* Salter). He makes allowance

TABLE 1.5. *Characteristics of the Advanced Capitalist Countries and the Rest of the World, 1900–1989*

	Per capita GDP: $ at 1985 US prices					Per capita GDP growth rate 1900–89	Population growth rate 1900–89
	1900	1913	1950	1973	1989		
Average for 16 advanced countries	2,374	2,937	4,693	10,396	14,456	2.1	0.9
Czechoslovakia	1,747	2,087	3,486	7,019	8,593	1.8	0.3
Hungary	1,617	1,883	2,481	5,517	6,598	1.6	0.4
New Zealand	3,090	4,106	7,577	10,156	11,067	1.4	1.6
Portugal	925	967	1,609	5,563	7,346	2.4	0.7
Spain	1,773	2,188	2,380	7,497	9,890	2.0	0.8
USSR	985	1,202	2,797	6,256	7,461	2.5	1.0
Middle income country average	1,690	2,072	3,388	7,001	8,493	2.0	0.8
Argentina	1,724	2,377	3,121	4,987	3,880	0.9	2.2
Brazil	586	700	1,441	3,363	4,241	2.2	2.4
Chile	1,284	1,685	3,156	4,444	5,355	1.6	1.7
Mexico	872	1,104	1,570	3,155	3,521	1.6	2.1
Latin American average	1,117	1,467	2,322	3,987	4,249	1.6	2.1

Bangladesh	469	498	445	377	504	0.1	1.5
China	539	557	454	1,039	2,361	1.7	1.2
India	508	536	482	689	1,065	0.8	1.4
Indonesia	670	710	650	1,056	1,790	1.1	1.7
South Korea	737	819	757	2,404	6,503	2.5	1.8
Taiwan	583	608	706	2,803	7,252	2.9	2.2
Asian average	584	621	582	1,395	3,246	1.5	1.6

Sources: Advanced capitalist countries from Appendices A and B. GDP and population growth in other countries from A. Maddison, *The World Economy in the Twentieth Century*, OECD, 1989, and A. Maddison, 'Measuring European Growth: The Core and the Periphery', in E. Aerts and N. Valerio (eds.), *Growth and Stagnation in the Mediterranean World*, Leuven University Press, 1990, updated where necessary. Benchmark GDP levels in New Zealand, Portugal, and Spain converted with 1985 Paasche PPPs supplied by Eurostat. For other countries the basic estimates of PPP were in 1980 international prices as indicated in Maddison, *World Economy*, 1989, except for Hungary, where the 1980 real product estimate was taken from *World Comparisons of Purchasing Power and Real Product for 1980*, UN/Eurostat, New York, 1986, p. 7. Czechoslovakia and the USSR were linked to the Hungarian level using ratios derivable from R. Heston and A. Summers, 'A New Set of International Comparisons', *Review of Income and Wealth*, Mar. 1988. The figures at 1980 prices were converted into 1985 US prices using a multiplier of 1.343, which was the average ratio of the 1985/1980 converters for the advanced countries. In all cases the group averages are arithmetic.

for gains owing to economies of scale, sectoral shifts in production structure, international specialization, and disembodied technical progress. He ends up with a measure of 'total factor productivity' and an unexplained residual.

All quantitative analysts of economic growth are greatly indebted to Denison, who has shown great ingenuity and sophistication in providing indications of the potential order of magnitude of particular influences on growth, and has demonstrated that respect for national accounting and its logic does not preclude intelligent guesswork. The rigour of his analysis, the meticulous detail of his research, and his analytical mastery have done a great deal to spark off further useful work in this field.[26]

Another major stream of post-war thought that is relevant to our interests is 'development economics', which deals with the growth problems of poor countries. Five main types of explanation for the lower income and productivity of 'developing' countries emerge from this literature: (1) the institutional setting was or is less favourable to capitalist development than that of western Europe and its offshoots; (2) various kinds of colonialism retarded development; (3) demographic growth has been much greater than was ever the case in the advanced capitalist countries; (4) their levels of investment in human and physical capital are very much lower than in the advanced countries; (5) inflationary and dirigiste policies have lowered efficiency.

As this study is concerned with advanced capitalist nations, it is not possible to analyse here the reasons for the divergent experience of middle-income and poor countries; nor is it possible to analyse the very different institutional arrangements and performance of Soviet-style economies. The per capita income range of our sixteen advanced countries is now quite narrow, with the worst-off having an income level only one third below that of the lead country — the USA. 'Developing' countries fall well outside this range. The poorest of them had an income only a fortieth of that in the USA in 1989 (see Table 1.5).

NOTES ON CHAPTER 1

1. I have in mind not only the defensive feudal order which emerged in Western Europe after the collapse of the Roman Empire, but also the characteristics of Asian societies; see the concept of 'high level equilibrium' in M. Elvin, *The Pattern of the Chinese Past*, Eyre Methuen, London, 1973, and D. Lal's interpretation of the origins and function of the caste system in India, in *The Hindu Equilibrium*, vol. i, Oxford, 1988.

2. The evidence (see Appendix A) that historical national accountants have built up over the past three decades has falsified the earlier view (see W. W. Rostow, *The Stages of Economic Growth*, Cambridge, 1960, p. 38) that there was a long-drawn-out sequence of staggered 'take-offs' in Western European countries throughout the 19th century. Recent scholarship, e.g. N. F. R. Crafts, *British Economic Growth during the Industrial Revolution*, Oxford, 1985, also suggests that 18th-century growth was slower than was previously thought. This recent evidence leads me to put the turning-point at 1820 rather than 1750 as Simon Kuznets did in his *Modern Economic Growth: Rate, Structure, and Spread*, Yale, 1966.

3. Apart from Kuznets, cited in Table 1.3, see D. S. Landes, *The Unbound Prometheus*, Cambridge, 1969, p. 14, who suggests that from the 11th to the 18th century, European real income per head may have tripled; C. M. Cipolla, *Before the Industrial Revolution: European Society and Economy, 1000–1700*, Norton, New York, 1976, also suggests a slow but rising long-run trend. For evidence of improvement in agricultural performance, see B. H. Slicher van Bath, *The Agrarian History of Western Europe AD 500–1850*, Arnold, London, 1963. For the increasing degree of urbanization, see J. de Vries, *European Urbanization 1500–1800*, Methuen, London, 1984.

4. Quantified propositions are easier to check and falsify than qualitative ones and hence they are useful in sharpening the debate on issues of comparative performance. Thus E. L. Jones, *Growth Recurring: Economic Change in World History*, Oxford, 1988, speaks of Sung China and Tokugawa Japan as if their performance were comparable to that of West European capitalist countries, but he cites little quantitative evidence to illustrate more specifically what he means. When he does use such data (for example the article by Susan Hanley, discussed in n. 6 to Appendix A below), it is easier to show that he is wrong.

5. See J. Needham, *Science and Civilisation in China*, vols. i–vi, Cambridge, 1954–84.

6. D. H Perkins, *Agricultural Development in China 1368–1968*, Aldine, Chicago, 1969.

7. See A. Maddison, 'Ultimate and Proximate Growth Causality: A Critique of Mancur Olson on the Rise and Decline of Nations', *Scandinavian Economic History Review*, No. 2, 1988.

8. T. R. Malthus, *First Essay on Population 1798*, Macmillan, London, 1966, p. 139.

9. For a discussion of the views of two leading neo-Malthusian historians (Emmanuel le Roy Ladurie and Wilhelm Abel), see A. Maddison, *Phases of Capitalist Development*, Oxford, 1982, pp. 11–12. Ladurie strongly emphasized the long-term stability of the French economy from 1300 to 1700,

both in demographic and per capita terms: see 'L'Histoire immobile', *Le Territoire de l'historien*, vol. ii, Gallimard, Paris, 1978, p. 16. Abel carried pessimism further, and suggested that real living standards in Germany and England actually fell from the first half of the 14th century to the first half of the 18th century: see *Agrarkrisen und Agrarkonjunktur*, Parey, Hamburg, 1978, pp. 285–9.

10. See P. Laslett, *The World We Have Lost*, Methuen, London, 1973, chs. 4 and 5. Laslett also describes the work of the Cambridge Group for the History of Population and Social Structure, which has been strongly influenced by Louis Henry of the Institut National d'Études Démographiques in Paris, and French historical demographers of the Annales school.

11. See G. E. Barnett (ed.), *Two Tracts by Gregory King*, Johns Hopkins, Baltimore, 1936, p. 31.

12. See E. Boserup, *The Conditions of Agricultural Growth*, Allen & Unwin, London, 1965, for a major contribution to anti-Malthusian analysis.

13. See L. White, *Medieval Technology and Social Change*, Oxford, 1962; B. H. Slicher van Bath, *The Agrarian History of Western Europe AD 500–1850*, Edward Arnold, London, 1963.

14. On the plunder element in leading countries of the early protocapitalist period, see B. L. Solow, 'Capitalism and Slavery in the Exceedingly Long Run', *Journal of Interdisciplinary History*, Spring 1987, pp. 711–37; for the Dutch experience, see A. Maddison, 'Dutch Income in and from Indonesia, 1700–1938', in A. Maddison and G. Prince, *Economic Growth in Indonesia, 1820–1940*, Foris, Dordrecht, 1989.

15. See Richard Rapp's examination of 17th-century business records in Venice, and his findings on the importance of inventories or raw materials and final products relative to fixed capital, in *Industry and Economic Decline in Seventeenth Century Venice*, Harvard, 1976, pp. 116–21.

16. J. A. Schumpeter's posthumous *History of Economic Analysis*, Oxford, 1954, is an encyclopaedic and subtle review of economic literature revealing his great detachment and generosity as a critic. Marx also left a much less polished manuscript survey, *Theories of Surplus Value* (3 vols.), Lawrence and Wishart, London, 1969, written in the opposite style with vituperative fervour.

17. See P. Sraffa and M. H. Dobb (eds.), *The Works and Correspondence of David Ricardo* (10 vols.), Cambridge, 1951. The most succinct presentation of Ricardo's schema is in his 'Essay on Profits', vol. iv, pp. 9–41, published in 1815.

18. Marx's vision of 'the laws of motion' of the capitalist epoch is fairly fully stated in chs. 22–4 of vol. i of *Capital*. His cycle theory and analysis of different types of unemployment is contained in sections 3 and 4 of ch. 23. For a critical assessment and reader's guide, see M. Blaug, *Economic Theory in Retrospect*, Heinemann, London, 1968, ch. 7. See also M. Dobb, *Political Economy and Capitalism*, Routledge & Kegan Paul, London, 1937, ch. V. My presentation of Marx is rather simplified and stresses elements with which I am largely in agreement, ignoring the labour theory of value and the division of capital into circulating and fixed. The problem in interpreting Marx is that his major published work on capitalist development was supplemented after his death by 4,500 pages of his preparatory or unfinished work on the same topic.

19. Here is what he said about the possibility of diminishing returns on land: 'Technological progress effectively turned the tables on any such tendency, and it is one of the safest predictions that in the calculable future we shall live in an *embarras de richesse* of both foodstuffs and raw materials, giving all the rein to expansion of total output that we shall know what to do with. This applies to

mineral resources as well'. See J. A. Schumpeter, *Capitalism, Socialism, and Democracy*, Allen & Unwin, London, 1943, p. 116.

20. See J. A. Schumpeter, *Imperialism, Social Classes* (ed. B. Hoselitz), Meridian, New York, 1951.

21. The above statement of Schumpeter's views is from *The Theory of Economic Development*, Oxford University Press, New York, 1961, which he first published in German in 1911. In later life, he expressed great detachment about the fate of capitalism and stressed that he had not intended to glorify entrepreneurs. In a note to the first English-language edition of his book, which appeared in 1934, he even suggested that their economic function could not be distinguished from that of robbers (p. 90). In a later work (see *Business Cycles*, 1939), he described how the entrepreneurial function can be institutionalized in large corporations, and also put forward a much more complex cycle theory.

22. See W. E. G. Salter, *Productivity and Technical Change*, Cambridge, 1960. Salter's approach to embodied technical progress is very similar to that of Robert Solow, who introduced the use of the term 'vintage' in this context: see R. M. Solow, 'Investment and Technical Progress', in K. J. Arrow, S. Karlin, and P. Suppes (eds.), *Mathematical Methods in the Social Sciences, 1959*, Stanford, 1960, pp. 89–104.

23. See T. W. Schultz, 'Investment in Human Capital', *American Economic Review*, Mar. 1961.

24. See A. Maddison, 'What is Education For?', *Lloyds Bank Review*, Apr. 1974, for an attempt to classify the different objectives of education and assess the contribution of the human capital school.

25. See E. F. Denison and J. P. Poullier, *Why Growth Rates Differ*, Brookings Institution, Washington, 1967, which deals with the situation in nine advanced capitalist countries from 1950 to 1962. He has also written four studies on US growth since 1929, of which the latest is *Trends in American Economic Growth 1929–1982*, Brookings Institution, Washington, 1985, and (with W. K. Chung) *How Japan's Economy Grew So Fast*, Brookings Institution, Washington, 1976. The 1967 study is the one to which I refer here.

26. See A. Maddison, 'Explaining Economic Growth', *Banca Nazionale del Lavoro Quarterly Review*, Sept. 1972, and A. Maddison, 'Growth and Slowdown in Advanced Capitalist Economies: Techniques of Quantitative Assessment', *Journal of Economic Literature*, June 1987, for a survey of the growth-accounting literature, and an explanation of why I give a greater weight to capital than Denison does.

2

CHANGES IN ECONOMIC LEADERSHIP

In analysing long-run development, it is important to distinguish between the 'lead' country, which operates nearest to the technical frontier, and 'follower' countries, which have a lower level of productivity. In the past four centuries there have been only three lead countries. The Netherlands was the top performer until the Napoleonic Wars, when the UK took over. The British lead lasted till around 1890, and the USA has been the lead country since then.[1] Leadership is defined here in terms of labour productivity (GDP per man-hour, see Graph 2.1).

It is useful to see why these countries attained their lead position, why the first two lost it, and whether the USA is now about to lose it. This should help to illuminate which factors are important in growth, and what the underlying growth potential is for the advanced capitalist group as a whole. The forces animating a lead country are more mysterious and autonomous than in the followers, whose growth path can be more easily influenced by policies to mimic the achievements of the leader and exploit the opportunities of relative backwardness.

THE DUTCH CASE

Around 1500, economic leadership was concentrated on Northern Italy and what is now Belgium. At that time, economic performance in the area which later became the Dutch republic was well below that in the southern part of the Low Countries, which had bigger towns, a larger textile industry, and much more international trade.

In the course of the sixteenth century there was a transformation of Dutch agriculture, very fast growth in the Dutch urban population and in international trade.[2] These developments were enhanced after the Netherlands achieved its independence and threw off the fiscal burdens levied by its former Spanish overlords.

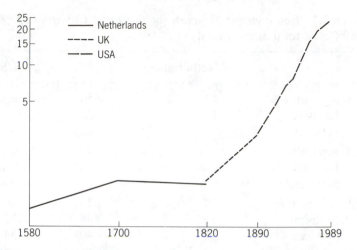

Graph 2.1 *Locus of Productivity Leadership, 1580–1989* (GDP per man-hour in 1985 US$)

A substantial proportion of the most enterprising and skilled population of Antwerp emigrated to Amsterdam, and it also benefited from the shift in world trade away from the Mediterranean to the Atlantic and other oceans.[3]

By 1700 Dutch income per head was around 50 per cent higher than that in the nearest rival, the UK, and its economic structure was more advanced. Only 40 per cent of employment was in agriculture as compared with 56 per cent in the UK (Table 2.1). The degree of urbanization was higher. International trade was as big as that of the UK although the Dutch population was only a fifth as large.

Dutch agriculture was highly specialized, with exports of dairy products, imports of live cattle from Denmark, and a quarter of grain needs supplied by Poland. Non-food crops such as hemp, flax, bulbs, hops, madder, and tobacco were of considerable significance. The service sector had developed large-scale international transactions in banking, insurance, shipping, and warehousing. There was a diversified industry, with sophisticated processing (bleaching, printing, and dyeing) of English woollens and German linens, as well as major domestic production of linen. Other significant industries were malting, brewing, distilling,

TABLE 2.1. *Employment Structure in the Lead Countries, 1700–1989* (% of total employment)

	Netherlands	UK	USA
1700 Agriculture	40	56	n.a.
Industry	33	22	n.a.
Services	27	22	n.a.
1820 Agriculture	n.a.	40	n.a.
Industry	n.a.	32	n.a.
Services	n.a.	28	n.a.
1890 Agriculture	33	16	39
Industry	31	44	27
Services	36	40	34
1989 Agriculture	5	2	3
Industry	26	29	26
Services	69	69	71

Note: Agriculture includes forestry and fishing; industry includes mining, manufacturing, electricity, gas, water, and construction; services is a residual including all other activity, private and governmental (including military).

Sources: For the Netherlands, the 1700 figures are an average of the proportions given for the 18th century for Friesland and Overijssel by Jan de Vries, *The Dutch Rural Economy in the Golden Age, 1500–1700*, Yale, 1974, p. 230; and Slicher van Bath, *Een Samenleving Onder Spanning*, Van Gorcum, Assen, 1957, p. 126, respectively. Both these areas had remarkably high industrialization for such relatively rural provinces. We know that western Holland with its big towns had even bigger proportions in industry and services, see A. M. van der Woude, *Het Noorderkwartier*, Wageningen, 1972, though the only occupational information given in the latter source is for 1811 (p. 270). Netherlands, 1890, derived from P. Bairoch, *The Working Population and its Structure*, Brussels, 1968. For the UK, 1700 and 1820, from P. H. Lindert and J. G. Williamson, 'Revising England's Social Tables 1688–1812', *Explorations in Economic History*, 19, 1982, with Irish structure assumed to be as in 1700. UK, 1890, derived from C. H. Feinstein, *National Income, Expenditure and Output of the United Kingdom 1855–1965*, Cambridge, 1972, p. T131. USA, 1890, derived from J. W. Kendrick, *Understanding Productivity*, Johns Hopkins, Baltimore, 1977, p. 43. 1989 figures for all countries from OECD, *Labour Force Statistics*, Paris.

ceramics, soap, bricks, tobacco cutting, tanning, sugar refining, shipbuilding, and fisheries. The fuel of Dutch industry was peat, which was available at low cost compared with energy prices in other countries, because it could be transported by canal. Dutch industries therefore tended to be energy-intensive.[4]

Capital formation was high by the standards of the epoch, with heavy investment in land reclamation, canals, urban infrastructure, windmills, shipping, and sawmills.[5] There was significant technological innovation in agriculture, ship design, shipbuilding techniques, saw milling hydraulic engineering, etc. Dutch universities had a high level of technical achievement, and the country had artisans skilled in making machinery, optical instruments, and clocks.

There were three main reasons for Dutch achievement. The first was the 'modernity' of Dutch institutions, which were highly favourable to capitalist enterprise, and the absence of a feudal past in most of the country.[6] Most land was owned by peasant proprietors or tenants with capitalist-type leases and money rents. There was relatively little common land to constrain cropping practice or inhibit efficient livestock breeding. There was a rather small nobility of landowners, and virtually no Church land after the departure of the Spanish. Political power resided largely in the urban bourgeoisie. All land was clearly registered and mortgageable. This meant greater freedom than elsewhere to vary output patterns in agriculture. It also meant that urban activity was relatively free of guild restrictions.

Enterprise was further helped by religious tolerance and immigration of Protestant and Jewish refugees. The unusual character of Dutch society was not a product of the Protestant ethic, but was due to the unique origins of the country. It was a country of recent settlement — on land mostly retrieved from the sea and marshes. Hence Dutch views were deeply impregnated with the possibilities for rational manipulation of the human and material environment and a 'Faustian sense of mastery over man and nature', which characterize capitalist attitudes to technological change.[7]

The second element that the Dutch turned to advantage was geography. The Netherlands dominated major rivers giving access to markets in the heart of Europe. There was scope for developing ocean ports, and possibilities for internal water transport were exploited to provide regular and frequent services for passengers and freight at lower cost than in any other country at that period.[8] They were short on timber, but turned peat into a major fuel, and used wind power imaginatively to operate drainage works and propel industrial machinery.

The third reason for success was the pursuit of mercantilist policies. The world was still a place with limited markets, in which the success of trading countries depended on beggar-your-neighbour practices. Thus the Netherlands blockaded Antwerp's access to the sea from 1585 to 1795, taking over its entrepôt trade and textile industry. Its successful struggle with Portugal was similarly responsible for Dutch monopolies in trade with large parts of Asia and Latin America. It should be noted that the bulk of Dutch trade (probably three-quarters) was of an entrepôt character, involving transhipment and warehousing in Amsterdam.

The main reason for loss of leadership was the destruction of monopolistic trading privileges in conflicts with France and the UK, which pushed the Dutch to the sidelines. Nevertheless, the economy did not collapse. It simply entered on a long period of decadence throughout the eighteenth century.

Population growth slackened as the economy ceased to attract migrants. The importance of international trade declined. There was some de-urbanization with a growth in the proportion employed in agriculture. There was stagnation in population in the industrialized western Netherlands and substantial growth in the agricultural province of Overijssel. Agricultural output increased, with a fall in imports and a growth in agricultural exports. There was a decline in production and exports of the leading industries of the seventeenth century. These included textiles (particularly the Leiden woollen industry), fisheries, and shipbuilding.[9] The volume of foreign trade dropped almost 20 per cent from 1700 to 1790 (before the French wars temporarily brought it to a standstill). During this period UK exports rose almost fourfold in volume, and French two-and-a-quarter-fold.

Dutch service industries continued to play an important part in the economy, and there was a large increase in overseas investment. In 1790 total foreign investment probably amounted to 800 million guilders at a time when national income was around 440 million. If the rate of return on foreign investment was around 4 per cent, then foreign income would have been around 30 million guilders, giving a national income about 8 per cent higher than domestic product.[10] The combination of rising *rentier* incomes, together with pauperism and unemployment in the old industrial areas, increased inequality.

There seems little doubt that a contributory factor to the eighteenth-century decline was that the currency was overvalued in the new international trading situation. There were numerous complaints about the high level of Dutch wages in the eighteenth century, and the huge efflux of capital speaks for itself. The exchange rate remained unchanged throughout 1700–75.[11]

Other explanations have been advanced for Dutch performance in the eighteenth century. It has been argued that the economy sank into complacent stagnation because of its 'high-level' traditionalism.[12] This argument involves the idea that commerce was favoured at the expense of manufacture, that the life-style was so satisfying and technology so close to the frontier of contemporary best practice that there was little incentive to innovate. It is true that performance was closer to potential than has been the case for most capitalist economies since then, but I do not see that this should lead to loss of entrepreneurial dynamism. My own hunch is that the primary cause was British and French damage to Dutch foreign markets, together with an overdevelopment of the banking interest which kept the currency overvalued and further weakened export potential.[13]

THE BRITISH CASE

At the end of the seventeenth and for a good part of the eighteenth century, British economic writers like Petty, Child, Temple, and Adam Smith advocated emulation of the Dutch model, in respect of institutions, technology, and the greater division of labour by international specialization.

In fact, most of British progress in the eighteenth century was a replication of the Dutch merchant capitalist model. British progress in this era depended heavily on Dutch technology. This was true in agriculture, canal building, shipping, banking, and international specialization. In agriculture and services, therefore, the UK was a follower, not a leader.[14] By 1820, when the UK became the lead country, its income per head and productivity were no better than the Netherlands had achieved in 1700 (see

Tables C.12 and C.13), and its economic structure was similar (see Table 2.1). The crucial difference between the UK in 1820 and the Netherlands in 1700 was the emergence of rapid technical progress in cotton textiles, iron manufacture, and the use of coal which the UK pioneered but which were still a small part of the economy.[15]

The reasons for British technological advance in the eighteenth century were rather similar to those of the Dutch in the seventeenth. These were (1) the character of British institutions, which permitted economic change and enterprise in agriculture, industry, and commerce and favoured the development of scientific attitudes;[16] (2) some of the geographic advantages of the Netherlands with regard to sea transport and the same kind of challenge-response mechanism with regard to energy: shortage of timber drove the Dutch to peat and the British to coal; (3) the fact that the UK took over the Dutch leadership as a world trader. It was this latter characteristic of the merchant capitalist situation which probably did most to launch the UK into the new textile technology, for it greatly expanded the market in this field.

The rise of British trading monopolies and their dominance of world trade was the decisive factor in Dutch decline. Thanks to mercantilist policy and a powerful navy, the British destroyed Dutch commercial supremacy in a series of wars at the end of the seventeenth century, and built up their own monopolies and privileges in North America, the West Indies, India, Brazil, and Spanish America (the latter by virtue of the Methuen and Utrecht treaties). British merchants had long experience as textile traders, both in exporting domestic woollens and importing or re-exporting Indian cottons. In 1760, when the wave of innovation in cotton textiles began, the UK was already the world's biggest trading country, and almost two-thirds of its exports were textiles. In textiles, the export market was as big as the internal market.

This emphasis on the mercantilist element in UK access to leadership may understate British technical virtuosity. However, the acceleration of technical progress before the nineteenth century was a gradual process, which covered a much wider area than textiles, which are usually given so much credit as a lead sector. It was the fruit of a cumulative process in Western civilization, requiring slow institutional change and scientific advance in a number of countries. The wide diffusion of the process is clearest in agriculture, but even in cotton textiles UK experience was not unique. Cotton textile output expanded rapidly in France as well as the UK, and the *ancien régime* in France did not prevent that country from experiencing significant economic development in the eighteenth century, albeit less than that in the UK.[17]

The textile innovations of the eighteenth century did not involve great scientific novelty or heavy investment, many of them were made by artisans, and in some respects it is a puzzle why they were not developed earlier. Nevertheless, they were a spectacular demonstration of the potential profits from innovation. Hargreaves's jenny (1764–7) permitted a sixteenfold productivity increase in spinning soft weft, and was an immediate success in the home spinning industry. Arkwright's spinning frame (1768) could produce a strong warp and used water power. Crompton's 1779 'mule' could produce both weft and warp. Cartwright's 1787 power loom extended the productivity gains to weaving; and, finally, the American Eli Whitney invented the cotton gin in 1793, which substantially reduced the cost of raw cotton production.

The highly profitable experience with textiles spread new attitudes to innovation throughout the economy. But the new wave of innovations in steam power, steel, and railways were of a different category from the textile inventions, depending more clearly on application of scientific principles and requiring much bigger investments for their application. They made a decisive change in the use of non-human power and in the capacity to transport goods to much bigger markets. The process of innovation was rapidly diffused to other countries which were institutionally ripe for capitalist development.

Although British nineteenth-century productivity growth was faster than it had ever been in the Netherlands, we know in retrospect from US experience that the UK was not a particularly dynamic leader. Its labour productivity grew from 1820 to 1890 at 1.2 per cent a year, whereas American productivity grew 2.2 per cent a year from 1890 to 1989. Joint factor productivity (output increase relative to the growth of combined inputs of labour and capital) grew by 0.9 per cent a year in the UK from 1820 to 1890, but in the USA it grew by 1.5 per cent a year from 1890 to 1989 (see Table 2.2).

The UK did little by way of government policy to foster either education or technical progress. An appreciable portion of its most

TABLE 2.2. *Growth of Productivity and Capital Stock in the UK and USA, 1700–1989*

Lead country	Date	Annual average compound growth rates		
		GDP per man-hour	Fixed non-residential gross capital stock per man-hour	Joint factor productivity
UK	1700–80	0.3	n.a.	n.a.
UK	1780–1820	0.4	0.1	0.4
UK	1820–90	1.2	1.6	0.9
USA	1890–1989	2.2	2.3	1.5

Sources: Appendix C, Tables C.10, C.11, and C.12; Appendix D, Table D.10. UK gross capital stock, 1780–1890, from C. H. Feinstein and S. Pollard, *Studies in Capital Formation in the United Kingdom 1750–1920*, Oxford, 1988, pp. 450–1. Joint factor productivity was calculated by giving labour input (total hours worked) a weight of 0.7 and capital input (gross nonresidential capital stock) a weight of 0.3. For a detailed discussion of the large variety of more complex joint factor productivity measures, see A. Maddison, 'Growth and Slowdown in Advanced Capitalist Economies: Techniques of Quantitative Assessment', *Journal of Economic Literature*, June 1987, and chapter 5 below.

ambitious people went into military bureaucratic posts overseas or were snobbish about entering domestic industry. Nor did it follow policies to stimulate high levels of domestic demand. Policy contributed to the high average level of productivity mainly by opening up the economy to international competition. This had obviously beneficial effects on productivity levels in a sector like agriculture, where British productivity was much higher than in those European countries which had high protective barriers. It also meant that the UK economic structure was more efficient because it concentrated on sectors in which it could perform well. Its service sector was highly developed, particularly the lucrative banking, shipping, and commercial services, which relied heavily on overseas markets and were helped both by the openness of the economy and the UK's position as an imperial power.

From 1820 to 1890 British capital stock per man-hour grew by 1.6 per cent a year. This was a great acceleration compared with earlier experience (see Table 2.2), but the UK did not push its investment effort as far as the USA was subsequently to do, and this is probably the major reason why it did not open up the technological frontier as fast. Growth in the domestic capital stock was not held up by a shortage of savings. In the decade before 1914, UK foreign investment was as big as domestic investment, and in 1914 its foreign assets were 1.5 times as high as its GDP.[18] In 1913, income from foreign property meant that GNP was 9.4 per cent higher than GDP. In 1855 the difference was only 2.1 per cent.

At the time it was overtaken by the USA, there were signs that the UK was growing at less than its potential because its currency was overvalued, it became a massive exporter of capital, and it experienced net emigration on a bigger scale than other European countries in spite of its higher income per capita. Its share of export markets declined. From 1820 to the mid-1850s, UK export prices had fallen more than import prices; from then onwards the external price movements were reversed. The check to profitability that overvaluation brought also inhibited domestic investment and labour productivity, which grew more slowly in 1890–1913 (1 per cent a year) than from 1820 to 1890.

The slackening of British productivity growth might have been alleviated by devaluation but the cost of the move would have been considered too damaging to the UK's interests as the manager of the world currency system, and to its banking and commercial interests. It was therefore never openly discussed. Even if such action had been taken, the UK would still have lost its leadership to the USA.

The British 'climacteric' at the latter third of the nineteenth century has been extensively discussed. However, the extent of the problem is sometimes exaggerated. The UK lost its productivity lead to the USA in the 1890s, but it kept ahead of European countries until well after the Second World War.[19] It did not lose it for the same reasons as the Netherlands did. It was not thrust out of a monopoly position in a limited world market; nor did it suffer an absolute decline in GDP. British loss of leadership was in fact inevitable, given the greater dynamism of the US economy.

From the viewpoint of the rest of the world, British policy was not unfavourable to growth. It diffused the growth process to follower countries by removing restrictions on trade, by investing abroad, and by permitting export of skills and technology, which was not the case in the merchant capitalist era.

THE US CASE

The emergence of the USA as the technical leader was due mainly to its large investment effort. The rate of US domestic investment was nearly twice the UK level for the sixty-year period 1890–1950 (see Table 2.3). Its level of capital stock per person employed was twice as high as that in the UK in 1890, and its overwhelming advantage in this respect over all other countries continued until the early 1980s (see Table 3.9). Between 1890 and 1913, US capital stock per person employed grew three times as fast as in the UK (see Table 5.6). The USA also had huge natural resources of land and minerals,[20] which by 1890 had been opened up by improvements in transport and the creation of a vast internal market whose population was much bigger than that of any of the advanced European countries, and was growing much faster due to immigration and high fertility.

At the time when leadership passed to the USA in 1890, American economic structure was less 'mature' than that of the UK (see Table 2.1). It had a much lower proportion of employment in industry, where its productivity was already nearly twice the UK level.[21]

TABLE 2.3. *Ratio of Gross Fixed Non-Residential Investment to GDP at Current Market Prices, 1871–1987*

	France	Germany	Japan	UK	USA[b]
1871–80	9.0	n.a.	n.a.	7.5	11.5
1881–90	10.4	n.a.	8.9[a]	5.9	12.2
1891–1900	10.4	n.a.	11.2	6.8	15.8
1901–10	10.4	n.a.	11.0	7.4	15.7
1911–20	n.a.	n.a.	14.9	6.2	12.5
1921–30	12.1	11.9	13.9	6.4	12.7
1931–40	11.1	10.1	14.7	6.5	9.7
1941–50	9.1	8.4	14.3	6.3	9.9
1951–60	13.8	16.6	20.1	12.4	12.6
1961–73	17.0	17.6	26.6	14.7	13.5
1974–87	14.9	14.6	23.9	14.1	13.6

[a] Refers to 1885–90.
[b] The first three entries refer to 1870–8, 1879–88, and 1889–1900, respectively.

Sources: France: 1871–1910 from M. Lévy-Leboyer and F. Bourguignon, *L'Économie française au XIXᵉ siècle*, Economica, Paris, 1985, pp. 330–2; 1922–38 from Carré, Dubois, and Malinvaud, *La Croissance française*, Seuil, Paris, 1972, p. 652. Both French sources were reduced 25 per cent to eliminate house construction and the repairs component of non-residential construction. 1921 and 1939–46 are my guesses. 1947 onwards from OEEC/OECD sources. Germany 1921–49 derived from W. Kirner, *Zeitreihen für das Anlagevermögen der Wirtschaftsbereiche in der Bundesrepublik Deutschland*, Duncker and Humblot, Berlin, 1968; 1950 onwards from OECD sources. Japan: 1885–1904 GDP and non-residential fixed investment from K. Ohkawa, N. Takamatsu, and Y. Yamamoto, *Estimates of Long-Term Economic Statistics of Japan since 1868*, vol. i. from *National Income*, Toyo Keizai Shinposha, 1974, pp. 186 and 200. 1905–61 K. Ohkawa and H. Rosovsky, *Japanese Economic Growth*, Oxford, 1973, pp. 278–9 and 290–1. 1962 onwards from *National Accounts of OECD Countries*. UK 1871–1920 capital formation from C. H. Feinstein and S. Pollard, *Studies in Capital Formation in the United Kingdom, 1750–1920*, Oxford, 1988, pp. 429–30. 1921–49 capital formation and 1871–49 GDP (modified 1871–1920 for revisions in capital formation) from C. H. Feinstein, *National Income, Expenditure and Output of the United Kingdom 1855–1965*, Cambridge, 1972. 1950 onwards from *National Accounts of OECD Countries*. USA to 1928 derived from J. W. Kendrick, *Productivity Trends in the United States*, Princeton, 1961, pp. 296–8; S. Kuznets, *Capital in the American Economy*, Princeton, 1961; R. Goldsmith, *A Study of Saving in the United States*, Princeton, 1955, pp. 619 and 623; and *Historical Statistics of the United States*, 1960 edn., pp. 379–80. 1929–60 from *The National Income and Product Accounts of the United States, 1929–1982, Statistical Tables*, US Dept. of Commerce. 1960 onwards from *National Accounts of OECD Countries*.

The US economy was big enough to breed giant corporations with professional management[22] and large research budgets, which helped to institutionalize the innovation process in a way in which the UK had never done. The USA strengthened its lead in this respect by building up research departments in its major universities, and making special provision in land grant colleges for agricultural research. It later attracted distinguished immigrants to its university faculties, particularly as a result of European wars. Recorded research and development spending in the USA rose from 0.2 per cent of GDP in 1921 to a peak of near 3 per cent in the mid-1960s.[23] These figures exaggerate growth to some extent because previously unrecorded efforts became institutionalized and there was a build-up of government research on military and space technology with limited economic applicability. Nevertheless, the growth was spectacular and dwarfed efforts elsewhere.

During UK leadership the big change was the switch in the production process from manufacture to machinofacture. Generally speaking, there was not much change in the nature of final consumer goods. In the US leadership period, the traditional pattern of consumption was transformed, and the lead country was active in product as well as process innovation. This involved developments in the techniques of salesmanship, market research, advertising, consumer credit, etc., which had little counterpart in the UK period, and in which the USA has excelled.

The relative standing of the USA as a lead country was greatly strengthened during the First and Second World Wars. Both of these stimulated demand and did almost no damage to capital assets. Both events were major catastrophes for European countries, as the Second World War was for Japan as well. They suffered not only physical damage to capital stock and manpower, but a narrowing of the scope for economies of scale through international trade. Thus the leader–follower gap among the advanced capitalist countries was swollen well beyond what it would have been in peacetime circumstances, in spite of the major slow-down in the growth of the US capital stock in 1913–50 (see Tables 5.4 to 5.7).

Normally the change of leadership from a slow UK to a more dynamic USA should have brought an appreciable acceleration of productivity growth within the advanced capitalist world. In fact, the average productivity growth rate did accelerate slightly in

TABLE 2.4. *Characteristics of Lead Country and Follower Country Performance in 1987 (USA = 100)*

	France	Germany	Japan	Netherlands	UK
GDP per man-hour	94	80	61	92	80
Educational level of labour force	85	71	83	73	80
Sex mix of employment	101	103	102	104	101
Non-residential gross capital stock per person employed	95	104	93	95	68
Modernity of capital stock	114	104	126	98	114
Land area per person employed	31	12	8	8	12
Energy consumption per person employed	57	62	39	74	52
R. & D. per person employed	71	79	71	64	61
Scale of domestic economy	17	19	40	4	17
Foreign trade per person employed	240	343	110	528	192

Sources: Derived mainly from our Statistical Appendices. Energy inputs from sources mentioned in Table 5.16. Foreign trade refers to sum of 1987 exports and imports from IMF, *International Financial Statistics*. R. & D. from OECD Science and Technology Directorate.

1913–50 in spite of two wars. Without them, the acceleration would probably have been much greater.

Since the Second World War the European countries and Japan have greatly reduced the productivity gap between themselves and the USA. Since 1973, US productivity growth has decelerated sharply. There has been some slow-down in other countries, too, but they have continued to converge on the USA. The gap in productivity between the USA and the follower countries is now smaller than it has been for the past century (see Table 3.4 below). Table 2.4 shows the distance between the USA and the major follower countries in terms of the major characteristics which affect productivity performance.

It is clear that the outstanding US advantage in terms of natural resources has faded considerably. Its minerals no longer have their pristine richness, and innovations in bulk transportation technology have given European countries and Japan access to even richer natural resources than those in the USA. The size of the US internal market is no longer such a unique asset, given the remarkable reduction of international trade barriers which has occurred in the post-war period. The follower countries have greatly reduced the gap in capital per head.

The USA is still the lead country in per capita GDP and productivity, when its output is measured (as it is in the study) in purchasing power parity rather than exchange rates, but if 1973–87 rates of convergence are maintained, France, the country nearest to US productivity levels, will have overtaken the USA by 1992, and Japan, which is presently the furthest from US levels, will catch up by the year 2012.[24]

NOTES ON CHAPTER 2

1. In defining productivity leadership, I have ignored the special case of Australia, whose impressive achievements before the First World War were due largely to natural resource advantages rather than to technical achievements and the stock of man-made capital. For similar reasons, I call the USA a technical leader in 1989 even though its real GDP per head was lower than that of Kuwait. It should be noted that Graph 2.1 shows average levels of labour productivity in the lead country. This is not the technical frontier as represented by 'best practice' firms in the lead country. Their performance will be well above average practice, because they have newer or larger capital per employee, better management, or more skilled workers.
2. On the rise of the Netherlands, see Jan de Vries, *The Dutch Rural Economy in the Golden Age*, Yale, 1974.
3. On the decline of the Southern (Spanish) Netherlands, see H. van der Wee (ed.), *The Rise and Decline of Urban Industries in Italy and the Low Countries*, Leuven, 1988. On Italian experience, there are differing views: C. M. Cipolla, *Before the Industrial Revolution*, Norton, New York, 1976, stresses absolute decline (pp. 236–44), whereas R. Rapp, *Industry and Economic Decline in Seventeenth Century Venice*, Harvard, 1966, suggests that in Venice, at least, the picture was rather one of stagnation.
4. See J. W. de Zeeuw, 'Peat and the Dutch Golden Age. The Historical Meaning of Energy-Attainability', *AAG Bijdragen*, 21, Wageningen, 1978, pp. 3–31.
5. Gregory King estimated the Dutch savings rate to have been over 11 per cent of national income in 1688 compared with 4 per cent in England and France: see G. E. Barnett (ed.), *Two Tracts by Gregory King*, Johns Hopkins, Baltimore, 1936.
6. See Jan de Vries, 'On the Modernity of the Dutch Republic', *Journal of Economic History*, 1973, pp. 191–202, and *The Dutch Rural Economy in the Golden Age*, Yale, 1974.
7. See D. S. Landes, *The Unbound Prometheus*, Cambridge, 1969, pp. 15–39, for an excellent analysis of the influence of attitudes to science and technology on capitalist development.
8. See Jan de Vries, 'Barges and Capitalism: Passenger Transportation in the Dutch Economy, 1632–1839', *AAG Bijdragen*, 21, Wageningen, 1978, pp. 33–40.
9. For 18th-century Dutch experience, see the excellent survey by Johan de Vries, *De economische achteruitgang der Republiek in de achttiende eeuw*, Leiden, 1968.
10. See J. C. Riley, *International Government Finance and the Amsterdam Capital Market 1740–1815*, Cambridge, 1980, pp. 16 and 84, who estimates 575 million guilders for loans to foreign governments, to which other loans to the private sector and direct investment (e.g. of the Dutch East India Company) should be added. Other authors have suggested higher investments, but Riley thinks they exaggerate. The estimate of national income is derived from J. L. van Zanden, 'De economie van Holland in de periode 1650–1805', *Bijdrage en Mededelingen Geschiedenis der Nederlanden*, 1987. See also A. Maddison, 'Dutch Income in and from Indonesia 1700–1938', in A. Maddison and G. Prince (eds.), *Economic Growth in Indonesia, 1820–1940*, Foris, Dordrecht, 1989.

11. See J. J. McCusker, *Money and Exchange in Europe and America 1600–1775*, Macmillan, London, 1978, pp. 59–60. This source also shows interest rates.
12. See Jan de Vries, *The Economy of Europe in an Age of Crisis 1600–1750*, Cambridge, 1976, pp. 251–2.
13. The de Vries type of argument was used by M. Elvin, *The Pattern of the Chinese Past*, Eyre Methuen, London, 1973, in a somewhat more developed form to explain the technical stagnation of China after having made such a brilliant and precocious start. Elvin calls it a 'high level equilibrium trap'.
14. This is very clear in agriculture. The main changes in British practice, sometimes called the agricultural revolution, were largely an adoption of already established Dutch techniques. These included (1) development of new root crops for animal fodder such as turnips, clover, and lucerne, which increased soil fertility and permitted greater intensity of land use; (2) improvements in livestock breeding, selection, and output; (3) greater use of additives to increase soil fertility, including animal manure, marl (clay) on thin soils, and lime on muddy soils. There was also a significant increase in the use of new high-calorie crops — maize and potatoes — originating in the New World. See J. D. Chambers and G. E. Mingay, *The Agricultural Revolution 1750–1880*, Batsford, London, 1966; and P. Bairoch, 'Agriculture and the Industrial Revolution 1700–1914', *Fontana Economic History of Europe*, vol. iii, 1973, for an account of these changes. Some of these technical developments, which permitted an increase in agricultural output per person, also led to increased labour input per person, with less land lying fallow and harder work in hitherto slack seasons, which was necessary with increased livestock holdings. Hence labour productivity did not increase as much as per capita income. See J. D. Chambers, 'Enclosure and the Labour Supply', *Economic History Review*, 1952/3, pp. 319–43; and C. P. Timmer, 'The Turnip, The New Husbandry and the English Agricultural Revolution', *Quarterly Journal of Economics*, 1969, p. 392, who suggest that the changes in British agriculture at this time increased land rather than labour productivity.
15. In 1820 the new cotton textiles and the iron industry represented less than 12 per cent of British GDP. See P. Deane, *The First Industrial Revolution*, Cambridge, 1965, pp. 88 and 108, who gives figures for Great Britain which I have adjusted to a UK basis.
16. See A. E. Musson and E. Robinson, *Science and Technology in the Industrial Revolution*, Manchester, 1969.
17. Between 1785 and 1820 French cotton consumption rose from 4,000 to 19,000 tons, and UK consumption from 8,200 to 54,000; i.e. per capita growth rates of 4.2 per cent a year in both cases. In both years French per capita consumption was about a quarter of that in the UK. See B. R. Mitchell, *European Historical Statistics 1750–1970*, Macmillan, London, 1975, pp. 427–8.
18. In 1914 UK foreign assets amounted to £3.8 billion: see H. Feis, *Europe The World's Banker 1870–1914*, Kelley, New York, 1961; and GDP at market prices to £2.5 billion.
19. There is a good deal of literature concerning UK rivalry with Germany before the First World War, because Germany had a bigger population and army. But German overall productivity was a good deal below that in the UK and this was probably true of agriculture, industry, and services taken separately. D. S. Landes, *The Unbound Prometheus*, Cambridge, 1969, and W. A. Lewis, *Growth and Fluctuations 1870–1913*, Allen & Unwin, London, 1978, are two of many authors stressing the German rivalry. D. N. McCloskey, *Economic Maturity and Entrepreneurial Decline, British Iron and Steel, 1870–1913*, Harvard, 1973, goes to the other extreme by virtually denying that the UK was

overtaken by the USA. He has been trapped into this position by measuring performance in terms of total factor productivity, which tends to conceal the fact that a major reason why US labour productivity grew faster than that of the UK was that its investment effort was bigger.

20. G. Wright, 'The Origins of American Industrial Success 1870–1940', *American Economic Review*, Sept. 1990, pp. 651–68, gives a useful elaboration of the importance and scale of the US natural resource advantage, and its role in the US achievement of economic leadership.
21. D. N. McCloskey (ed.), *Essays in a Mature Economy: Britain After 1840*, Methuen, London, 1971, pp. 291–5, shows labour productivity in British coal-mining in 1907 at about half the US 1909 level; E. H. Phelps Brown, *A Century of Pay*, Macmillan, London, 1968, p. 59, cites further evidence for eight industries for 1907–9, with UK output per man averaging 52 per cent of that in the USA. Y. Hayami and V. W. Ruttan, *Agricultural Development; An International Perspective*, Johns Hopkins, Baltimore, 1971, pp. 327 and 331, show British agricultural output per male worker 13 per cent higher than that in the USA in 1890.
22. For a comparison of American, British, and German managerial organization over the past century, see A. D. Chandler, jnr., *Scale and Scope: The Dynamics of Industrial Capitalism*, Harvard, 1990. This study tends to neglect the greater US physical capital per worker, better natural resources, and higher levels of education. As a consequence, it exaggerates the role of economies of scale and managerial organization in explaining variance in overall productivity performance between the USA and UK.
23. See R. R. Nelson, M. J. Peck, and E. D. Kalachek, *Technology, Economic Growth and Public Policy*, Brookings Institution, Washington, 1967; and Z. Griliches, 'R and D and the Productivity Slowdown', *American Economic Review*, May 1980, pp. 343–7.
24. See D. Pilat, 'Levels of Real Output and Labour Productivity by Industry of Origin: A Comparison of Japan and the United States, 1975 and 1970–1987', *Research Memorandum*, No. 408, Institute of Economic Research, University of Groningen, 1991, for a detailed comparison, by industry of origin, of PPPs, real output, and productivity in these two countries. Although Japan is probably closer to US levels of productivity in manufacturing than is the case in European countries, its agricultural performance is abysmal, and its service productivity is rather poor.

3

LONG-RUN DYNAMIC FORCES IN CAPITALIST DEVELOPMENT

This chapter concentrates on some of the longer-term features of the capitalist epoch which help to explain or are associated with its dynamic character. It works through the quantitative evidence underlying the explanatory schema set out in Chapter 1 (Table 1.4) and sets out other distinguishing features which are relevant in understanding the long-term growth experience.

Since the Napoleonic Wars ended, all our countries, with the exception of Japan, which was a late-starter, have been involved in a process of substantial and sustained growth. The average GDP growth for 1820–1989 was 2.7 per cent a year. GDP per capita grew 1.6 per cent a year — about eight times as fast as in the protocapitalist epoch.

There are some gaps in the growth record between 1820 and 1870, which force us to concentrate on experience since the latter date, but the evidence available suggests clearly that 1820–70 performance had greater similarity to that for 1870–1913 than to experience before 1820.[1]

Apart from this central fact of rapid economic progress, there are two other major characteristics which need to be noted and explained. The first is that growth has varied significantly, as Tables 3.1, 3.2, and 3.3 show clearly. In particular there was a golden age of greatly accelerated growth in the quarter-century after the Second World War, and a very substantial deterioration after 1973, though performance in this last phase has been better than in any earlier period except the golden age. Analysis of this variation in growth momentum and the reasons for it can be found in Chapters 4, 5, and 6.

A third major characteristic of capitalist development has been the post-war convergence in levels of per capita income and productivity between these advanced countries. This is something which is not true of the world as a whole, so it is of particular interest.

TABLE 3.1. *Growth Rates of Real GDP per Head of Population, 1820–1989* (annual average compound growth rates)

	1820–70	1870–1913	1913–50	1950–73	1973–89	1820–1989	1870–1989
Australia	1.9	0.9	0.7	2.4	1.7	1.4	1.2
Austria	0.6	1.5	0.2	4.9	2.4	1.5	1.8
Belgium	1.4	1.0	0.7	3.5	2.0	1.5	1.5
Canada	n.a.	2.3	1.5	2.9	2.5	n.a.	2.2
Denmark	0.9	1.6	1.5	3.1	1.6	1.6	1.8
Finland	0.8	1.4	1.9	4.3	2.7	1.8	2.3
France	0.8	1.3	1.1	4.0	1.8	1.5	1.8
Germany	0.7	1.6	0.7	4.9	2.1	1.6	2.0
Italy	0.4	1.3	0.8	5.0	2.6	1.5	2.0
Japan	0.1	1.4	0.9	8.0	3.1	1.9	2.7
Netherlands	0.9	1.0	1.1	3.4	1.4	1.4	1.5
Norway	0.7	1.3	2.1	3.2	3.6	1.8	2.2
Sweden	0.7	1.5	2.1	3.3	1.8	1.6	2.1
Switzerland	n.a.	1.2	2.1	3.1	1.0	n.a.	1.8
UK	1.2	1.0	0.8	2.5	1.8	1.3	1.4
USA	1.5	1.8	1.6	2.2	1.6	1.7	1.8
Arithmetic average	0.9	1.4	1.2	3.8	2.1	1.6	1.9

Sources: Appendices A and B. The figures are adjusted to exclude the impact of boundary changes.

TABLE 3.2. *Growth Rates of Real GDP, 1820–1989* (annual average compound growth rates)

	1820–70	1870–1913	1913–50	1950–73	1973–89	1820–1989	1870–1989
Australia	10.1	3.5	2.2	4.7	3.1	5.2	3.3
Austria	1.4	2.4	0.2	5.3	2.4	2.0	2.3
Belgium	2.2	2.0	1.0	4.1	2.1	2.2	2.1
Canada	n.a.	4.1	3.1	5.1	3.6	n.a.	3.9
Denmark	1.9	2.7	2.5	3.8	1.7	2.5	2.7
Finland	1.6	2.7	2.7	4.9	3.1	2.7	3.2
France	1.2	1.5	1.1	5.0	2.3	1.9	2.2
Germany	1.6	2.8	1.3	5.9	2.1	2.5	2.8
Italy	1.2	1.9	1.5	5.6	2.9	2.2	2.6
Japan	0.3	2.3	2.2	9.3	3.9	2.8	3.8
Netherlands	1.8	2.3	2.4	4.7	2.0	2.5	2.8
Norway	1.8	2.1	2.9	4.1	4.0	2.7	3.0
Sweden	1.6	2.2	2.7	4.0	2.0	2.4	2.7
Switzerland	n.a.	2.1	2.6	4.5	1.3	n.a.	2.6
UK	2.0	1.9	1.3	3.0	2.0	2.0	1.9
USA	4.5	3.9	2.8	3.6	2.7	3.7	3.4
Arithmetic average	2.4	2.5	2.0	4.9	2.6	2.7	2.8

Source: Appendix A. The figures are adjusted to exclude the impact of boundary changes.

TABLE 3.3. *Phases of Productivity Growth (GDP per Man-Hour), 1870–1987* (annual average compound growth rate)

	1870–1913	1913–50	1950–73	1973–87	1870–1987
Australia	1.1	1.5	2.7	1.8	1.6
Austria	1.8	0.9	5.9	2.7	2.4
Belgium	1.2	1.4	4.4	3.0	2.1
Canada	2.3	2.4	2.9	1.8	2.4
Denmark	1.9	1.6	4.1	1.6	2.2
Finland	1.8	2.3	5.2	2.2	2.7
France	1.6	1.9	5.0	3.2	2.6
Germany	1.9	1.0	5.9	2.6	2.5
Italy	1.7	2.0	5.8	2.6	2.7
Japan	1.9	1.8	7.6	3.5	3.2
Netherlands	1.3	1.3	4.8	2.4	2.1
Norway	1.6	2.5	4.2	3.5	2.6
Sweden	1.7	2.8	4.4	1.6	2.6
Switzerland	1.5	2.7	3.3	1.2	2.2
UK	1.2	1.6	3.2	2.3	1.9
USA	1.9	2.4	2.5	1.0	2.1
Arithmetic average	1.7	1.9	4.5	2.3	2.4

Source: Appendix C.

Per capita real income performance ranged from 1.3 per cent a year for the UK to 1.9 per cent for Japan for 1820 to 1989. The UK, the pioneer in accelerated growth, has now been overtaken by several of the other countries; whereas Japan, whose modern growth did not start until the Meiji reforms of 1867, has shown the fastest growth, and now ranks fifth in per capita income. Finland, the second poorest country in 1820, has had the second highest growth rate (behind Japan), and the Netherlands, the second richest in 1820, has had the second slowest growth rate. Australia, the richest country in 1870, has had the slowest growth since then.

The convergency process has certainly not been monotonic. The per capita income spread between these countries (see Table 1.1) was 2.4 : 1 in 1820, 5.3 : 1 in 1870, and 5.5 : 1 in 1950, i.e. there was strong divergence over a period of 130 years. The convergency process has been concentrated on the post-war period. It has been so strong that the spread is now less than 1.5 : 1. The same convergence can be observed in Table 3.4 for levels of labour productivity. In 1870 there was a range of 7 : 1; in 1987 it was about 1.6 : 1. The dynamics of divergence and convergence have been complex and were disturbed by wars, system shocks, and business cycles. This comes out clearly in Graph 3.1, which presents binary comparisons of real per capita income growth on an annual basis for 1890–1989. All the country graphs are on the same scale so the long-term dimensions of growth are clear and comparable. Each graph compares performance in the lead country, the USA, and one of five follower countries.

The reasons for acceleration, slow-down, and convergence are analysed in detail in Chapter 5.

INSTITUTIONS

Capitalist development after 1820 was characterized by processes of accumulation, innovation, diffusion of technology, and personal enrichment which, by previous historical experience, were unprecedented in scope. These features required societal, intellectual, and institutional changes that had taken place over the preceding four centuries. These were complex and gradual, and most were not present in other parts of the world.

The most fundamental was the recognition of human capacity to

TABLE 3.4. *Comparative Levels of Productivity, 1870–1987* (US GDP per man-hour = 100)

	1870	1913	1950	1973	1987
Australia	132	93	67	70	78
Austria	51	48	27	59	74
Belgium	82	61	42	64	86
Canada	64	75	75	83	92
Denmark	59	58	43	63	68
Finland	34	33	31	57	67
France	56	48	40	70	94
Germany	50	50	30	64	80
Italy	41	37	31	64	79
Japan	19	18	15	46	61
Netherlands	88	69	46	77	92
Norway	48	43	43	64	90
Sweden	47	44	49	76	82
Switzerland	62	51	56	67	68
UK	104	78	57	67	80
USA	100	100	100	100	100
Arithmetic average of 15 countries (excluding USA)	62	54	43	66	79

Source: Appendix C.

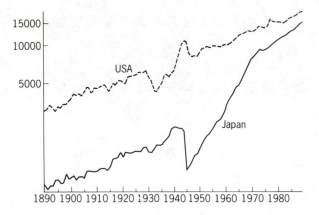

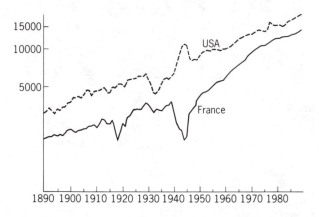

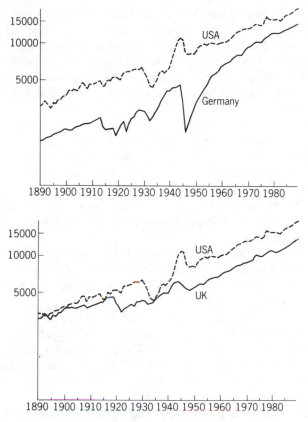

Graph 3.1 *Binary Comparisons of Per Capita GDP Growth, 1890–1989*
Vertical axis is logarithmic and shows level of GDP per capita in 1985
US$. Horizontal axis is chronological and marks decade intervals. Data
are annual.

transform the forces of nature through rational investigation and
experiment. Thanks to the Renaissance and the Enlightenment,
Western science abandoned superstition, magic, subordination to
religious authority, and circumscribed horizons.

The ending of feudal constraints on the free purchase and sale of
property was followed by a whole series of developments which
gave scope for successful entrepreneurship. A non-discretionary
legal system protected property rights. The development of

accountancy helped further in making contracts enforceable. State fiscal levies became more predictable and less arbitrary. The growth of trustworthy financial institutions and instruments gave access to credit and insurance which made it easier to assess risk and organize business rationally on a large scale over a wide area. Techniques of organization, management, and labour discipline grew.

A third distinctive feature of Western Europe was the emergence of a system of nation states in close propinquity, which had significant trading relations and relatively easy intellectual interchange in spite of their linguistic and cultural differences. This stimulated competition and innovation. Migration to or refuge in a different culture and environment were options open to adventurous minds; printing presses and universities added to the ease of interchange.

The Western family system was different from that in other parts of the world. It involved voluntaristic controls over fertility and limited obligations to more distant kin, which reinforced the possibilities for accumulation.

These societal features were very clear at an early stage in the three successive lead countries surveyed in Chapter 2. By 1820, they characterized all of the countries under consideration except Japan, which westernized its institutions in the Meiji reforms of 1867 and already had demographic characteristics closer to those of Europe than to the rest of Asia.

Since 1820, the institutional arrangements of advanced capitalist countries have not stood still. The degree of democratic participation and the socio-economic role of government have changed a good deal. In the post-war period, interrelations between these countries have involved articulate co-operation and some rudiments of a managed international order. These changes have affected the performance of these economies as noted at the end of this chapter and in Chapter 6.

NATURAL RESOURCES

We saw in Chapter 1 that there is a wide range of views about the limiting role of natural resources as a constraint on economic growth. These vary from the extreme pessimism of Malthus to the insouciance of Schumpeter.

TABLE 3.5. *Total Area per Head of Population* (hectares per person)

	1820	1987		1820	1987
Australia	2,308.36	47.27	Italy	1.59	0.53
Austria	2.82	1.11	Japan	1.31	0.31
Belgium	0.97	0.34	Netherlands	1.59	0.25
Canada	608.30	38.90	Norway	33.39	7.74
Denmark	3.93	0.84	Sweden	17.41	5.36
Finland	28.92	6.86	Switzerland	2.26	0.63
France	1.75	0.99	UK	1.48	0.43
Germany	2.17	0.41	USA	38.25	3.84

Sources: 1987 surface area from FAO, *Production Yearbook 1988*, Rome. For France, Germany, UK, and USA, 1820 area from national sources; for other countries, 1820 based on 1987 area.

The long-run experience since 1820 suggests that Malthus was certainly wrong and Schumpeter was too cheerful, though he was nearer to the truth.

Table 3.5 shows the total area per head of population in each country in 1820 and 1987. Surface area is only a crude proxy for natural resources, but it is clear (*a*) that natural resources per head declined substantially in all cases, and (*b*) that there are huge differences in endowment, with Australia, Canada, the USA, and Scandinavia having a good deal more land per head than the rest of Europe or Japan. The resource advantage of Australia, Canada, and the USA had a considerable influence on their total GDP growth, as it attracted a large flow of immigrants, but its influence on per capita GDP levels and growth has been decreasing over the long run. It is also clear that present real income levels have little relation to natural resource endowment. Per capita GDP in Japan is higher than in Australia even though Australian natural resources per capita are over 150 times greater.

The stock of agricultural land under cultivation in these countries grew in the nineteenth and first half of the twentieth century. But when this process ended, agricultural productivity grew faster than ever before, and has indeed grown faster than that in industry in virtually all these countries since 1950. Advances in technology and augmentation of land productivity by combining it with complementary reproducible capital have overcome the problem that Ricardo feared. The terms of trade of farmers have not improved but worsened. In most advanced countries governments help prop up their incomes by various devices.[2]

Although Malthusian fears gradually faded with regard to agricultural land, there has been recurrent concern about other natural resources. In 1865, the English economist Jevons[3] predicted that a coal shortage would bring economic growth to a halt within a century because there would be no substitute source of energy. Jevons did not foresee how many coal reserves would be discovered or how good a substitute oil would be. But even after this was demonstrated, conservationists continued to be gloomy about resource constraints and waste of non-renewable assets.

In the 1950s Colin Clark was worried about exhausting water resources, and the US government set up the Paley Commission because it was concerned about the adequacy of mineral reserves. In the 1970s the fear that economic growth was running into constraints because of the exhaustion of fixed natural resources was reiterated by the Club of Rome.[4] The sudden shocks caused by the oil price rises of the 1970s suggested that natural resources might check growth prospects, but as Table 3.6 shows, in energy, as with other resources, there has been a significant substitution between different sources, and a long-run decline in consumption per unit of real GDP.

In fact, prospectors usually discover new natural resources when prices rise, and technological progress has been rather successful in finding substitutes. In important cases, newly discovered resources have been richer than those discovered earlier. Rising prices check consumption and induce a switch to substitute products. Except in warlike situations, there are usually clear warnings of emerging scarcity as reflected in relative prices, and it is the rising relative price that induces technical change. Unfortunately, the price mechanism has been much less operative with regard to atmospheric and water pollution, as polluters have generally managed to avoid paying for the damage they do.

Past experience in successful adaptation to resource scarcity either domestically or by importation makes one sceptical about

TABLE 3.6. *The Pattern of US Primary Energy Requirements, 1890–1989*

	Total energy requirements per $1,000 of GDP at 1985 prices	Percentage derived from different sources					Total
		Wood	Coal	Oil and gas	Wind and water	Nuclear	
1890	0.90	35.3	57.0	7.5	0.2	0.0	100.0
1950	0.63	2.9	38.8	57.2	1.1	0.0	100.0
1989	0.43	0.8	26.2	63.5	3.1	6.4	100.0

Sources: 1989 from International Energy Agency, Paris. 1890–1950 from J. F. Dewhurst and Associates, *America's Needs and Resources*, Twentieth Century Fund, New York, 1955, p. 1114. Total energy requirements are in terms of metric tons of oil equivalent. Dewhurst shows them in short tons of coal equivalent. His figures were divided by 1.54 to convert to oil equivalent.

the likelihood of resource constraints on growth in the long run. However, it is clear that the size of the original endowment and the processes of resource exhaustion and discovery affect the fate of particular countries. Geography and natural resources favoured Dutch and British accession to economic leadership. The USA had great natural advantages in land, minerals, and energy resources which played a role in its subsequent leadership.

DEMOGRAPHIC CHANGE

Since 1820, population of these countries has increased 0.9 per cent a year, compared with 0.4 per cent in the eighteenth century. The acceleration took place despite a major decline in fertility. Birth-rates in the 1980s were only a third of those in 1820 and earlier. The increase was due to the prolongation of life expectation from 37 to over 76 years (see Table B.6).

Swedish experience of mortality is one of the best-documented cases, and is reasonably representative of the long-term development in the European countries. Graph 3.2 shows the gradual decline in 'normal' Swedish mortality from 25 per 1,000 in the eighteenth century to around 10 per 1,000 in the 1980s as well as the disappearance of periodic calamities caused by famine and disease which produced recurrent peaks in the mortality curve (and also reduced fertility by causing amenorrhoea). The last of these was the peak in mortality during the influenza epidemic of 1919. The general fertility experience has been like the Swedish, with a gradual and large long-run decline. The main difference between these countries has been in migration, which made population growth much faster in Australia, Canada, and the USA than in the European countries and Japan.

Demographic experience has been different from that in other parts of the world. In our group of countries population growth dropped to 0.5 per cent a year for 1973–89, compared with over 2 per cent in Asia and Latin America. The fall in mortality also occurred in Asia, Africa, and Latin America, though it was concentrated more heavily on the period since the Second World War. The really big difference is in fertility, which is very much lower in our countries than in Asia, Africa, or Latin America. The fertility differential has some relation to levels of income and

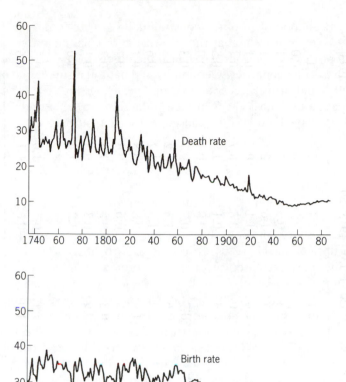

Graph 3.2 *Swedish Vital Statistics 1736–1987*
Vertical axis represents deaths and births respectively, per 1,000
population. Horizontal figures refer to 20-year intervals from 1740 to 1980
(sources: H. Gille, 'The Demographic History of the Northern Countries
in the Eighteenth Century', *Population Studies*, June 1949; *Historical
Statistics for Sweden*, vol. i, CBS, Stockholm, 1955; and OECD *Labour
Force Statistics*, Paris, various issues).

TABLE 3.7. *Rates of Population Growth, 1500–1989* (annual average compound growth rates)

	1500–1700	1700–1820	1820–1989	1820–1913	1913–50	1950–73	1973–89
Australia	0.00	0.1	2.3	3.1	1.4	2.2	1.4
Austria	0.20	0.3	0.5	0.9	0.1	0.4	0.0
Belgium	0.14	0.4	0.6	0.8	0.3	0.5	0.1
Canada	0.01	0.8	1.7	1.7	1.6	2.1	1.1
Denmark	0.08	0.4	0.9	1.0	1.1	0.7	0.1
Finland	0.14	0.2	0.9	1.0	0.8	0.7	0.4
France	0.13	0.3	0.4	0.3	0.1	1.0	0.5
Germany	0.11	0.4	0.5	1.1	−0.8	0.9	0.0
Italy	0.14	0.3	0.7	0.7	0.7	0.7	0.3
Japan	0.46	0.0	0.9	0.6	1.3	1.2	0.8
Netherlands	0.35	0.2	1.1	1.0	1.3	1.2	0.6
Norway	0.26	0.6	0.9	1.0	0.8	0.8	0.4
Sweden	0.42	0.6	0.7	0.8	0.6	0.6	0.2
Switzerland	0.20	0.4	0.8	0.8	0.5	1.4	0.2
UK	0.37	0.7	0.6	0.9	0.3	0.5	0.1
USA	0.05	1.2	1.8	2.3	1.2	1.4	1.0
Arithmetic average	0.19	0.4	1.0	1.1	0.7	1.0	0.5
Total	0.24	0.4	0.9	1.0	0.6	1.1	0.6

Source: Appendix B. The estimates are not adjusted for changes in frontiers except for Austria. The figures for Australia, Canada, and the USA include an estimate of the indigenous populations.

education, but to an important degree it reflects differences in societal attitudes which have led to later marriage and greater voluntaristic checks on the birth-rate in the West.

GROWTH OF LABOUR SUPPLY

Changes in demographic structure have led the labour force to increase faster than population in most of our countries.

The child proportion (below 15 years) decreased from a third of the population in 1870 to a fifth in the 1980s. This more than offset the substantial increase in people aged 65 and over, so that the population of working age (15–64) rose from 60 to 67 per cent of the total.

There was a decline in activity rates for males as opportunities for education and retirement income increased, but this was more than offset by the rise in female participation. By 1987, 60 per cent were in the labour force, which is probably more than twice the proportion in 1870.

The first decades of capitalist development involved an increase in working hours per person. Technical changes in agriculture were labour intensive, and working hours in industry were very long. In fact, they appear to have been irrationally long from the point of view of efficiency. It was not until the 1860s that this was recognized and that hours began to fall. Before that, factory work often involved a ten-hour day or more, and employers did not accept the argument that shorter hours would increase worker efficiency. Nassau Senior, the professor of economics at Oxford, defended their case by arguments which assumed that there was no efficiency gain from shorter hours.[5]

Since 1870, reduction in working hours and increase in vacations and time off for sickness have reduced the working year in these countries from 3,000 to an average of 1,600. Shorter hours improved the health of the labour force and increased the quality of labour input.

HUMAN CAPITAL

A significant characteristic of the advanced capitalist countries is the effort they have made over the long run to raise the level of

TABLE 3.8. *Average Years of Formal Educational Experience of the Population Aged 15–64 in 1913 and 1989*

		Total	Primary	Secondary	Higher
France	1913	6.18	4.31	1.77	0.10
	1989	11.61	5.00	5.29	1.32
Germany	1913	6.94	3.50	3.35	0.09
	1989	9.58	4.00	5.20	0.38
Japan	1913	5.10	4.50	0.56	0.04
	1989	11.66	6.00	4.95	0.71
Netherlands	1913	6.05	5.30	0.64	0.11
	1989	10.51	6.00	3.82	0.69
UK	1913	7.28	5.30	1.90	0.08
	1989	11.28	6.00	4.75	0.53
USA	1913	6.93	4.90	1.83	0.20
	1989	13.39	6.00	5.72	1.67

Sources: Generally from post-war census material, with estimates for 1913 extrapolated from information on education of older cohorts and for 1989 on information for younger cohorts. See A. Maddison, 'Growth and Slowdown in Advanced Capitalist Economies: Techniques of Quantitative Assessment', *Journal of Economic Literature*, June 1987, Table A12. USA 1989 extrapolated from E. F. Denison, *Trends in American Economic Growth 1929–82*, Brookings, Washington, DC, 1985, p. 91. These figures refer to full-time formal schooling and understate educational levels in Germany, which has an extensive system of post-formal apprentice training, combined with part-time education.

education of their populations. In 1820, their average education level for both sexes combined was probably about 2 years[6] and by 1989 this had risen to an average of over 11 years (see Table 3.8). Furthermore, education is now more evenly spread throughout the population thanks to universal attendance in primary and part of secondary education. The higher the average level of education, the easier it is for a working population to understand and apply the fruits of technical progress. It is difficult to be at all precise about the impact of rising educational standards on productivity, but most growth analysts consider it to have been substantial,[7] and it is striking that levels of education are so strongly related to the economic distance between nations. Within our sample, differences in income and levels of education are now not too large by international standards, but the gap between them and the poorer countries of the world is very large, as is the difference in incomes. In 1950, average educational levels in India were well below those in Europe in the 1820s, as were its income and productivity levels. Furthermore, the quality of education is generally better in these advanced nations than in most of the poorer countries.

PHYSICAL CAPITAL

There are three main kinds of physical capital: non-residential, residential, and inventories. The first is the biggest and has the greatest influence on growth; the second is substantial, but much more limited in its impact. Inventories are relatively small — about a third of GDP — and do not play an important explanatory role in long-term growth causality.

A necessary condition for exploiting the possibilities offered by technical progress is an increase in the stock of machinery and equipment in which this technology is embodied, and the buildings and infrastructure in which they operate. Table 3.9 shows the impressive growth of non-residential gross capital stock per person employed. The UK, which was the economic leader until 1890 and has since had the slowest productivity growth of the countries listed, had the slowest growth of capital per person employed (2.1 per cent a year since 1890). Conversely, Japanese capital stock per employee was initially the lowest and has grown the fastest (4.2 per cent a year since 1890), which is equally true of Japanese productivity experience. One can also see that the US non-residential capital stock per employee was more than double that of the UK in 1890, when it overtook the UK in terms of productivity. US productivity leadership over the past century has been accompanied throughout by a superior level of capital; but by the late 1980s, when its productivity lead had narrowed, it had begun to lose its marked superiority.

There has been a broad similarity in the *phasing* of growth rates

TABLE 3.9. *Gross Non-Residential Fixed Capital Stock per Person Employed, 1890–1987* ($ at 1985 US relative prices)

	1890	1913	1950	1973	1987
France	n.a.	(9,600)	14,800	43,309	80,604
Germany	(9,611)	(13,483)	16,291	55,421	89,154
Japan	1,454	2,264	6,609	33,101	78,681
Netherlands	n.a.	n.a.	20,181	59,459	80,897
UK	7,634	9,780	13,923	39,100	58,139
USA	16,402	35,485	48,118	70,677	85,023

Sources: Gross capital stock (non-residential structures and equipment) and employment estimates derived from sources indicated in Appendices C and D. For Germany 1890–1950 and France 1913–50 standardized gross capital stock estimates were not available so net stock movements were used as a proxy from national sources cited in A. Maddison 'Growth and Slowdown in Advanced Capitalist Economies', *Journal of Economic Literature*, June 1987, Table A.16. Capital stocks in national currencies were revalued at 1985 prices using implicit deflators for the relevant types of investment. Conversion into dollars was made by using 1985 purchasing power parities for investment goods supplied by Eurostat. The valuation basis is therefore comparable with that used for GDP.

for capital and output with a slackening in the 1913–50 period and an unprecedented acceleration in the post-war golden age (see Table 5.6). However, since 1890, there has been a very substantial increase in the non-residential capital output ratio in the follower countries, and particularly in the post-war period.[8] This is clear from Table 3.10 and in Graph 3.3, where the rise in the Japanese capital/output ratio is very clear.

TECHNICAL PROGRESS

Technical progress is the most essential characteristic of modern growth and the one that is most difficult to quantify or explain. Its effects are diffused throughout the growth process in a myriad ways. It augments the quality of natural resources and human

TABLE 3.10. *Ratio of Gross Non-Residential Capital Stock to GDP, 1890–1987* (at 1985 US relative prices)

	1890	1913	1950	1973	1987
France	n.a.	(1.64)	1.68	1.75	2.41
Germany	(2.29)	(2.25)	2.07	2.39	2.99
Japan	0.91	1.01	1.80	1.73	2.77
Netherlands	n.a.	n.a.	1.75	2.22	2.74
UK	0.95	1.03	1.10	1.73	2.02
Average excluding USA	1.38	1.48	1.68	1.96	2.59
USA	2.09	2.91	2.26	2.07	2.30

Sources: Appendices A and D. For bracketed figures see note to Table 3.9.

capital, and the impact of foreign trade. But investment is the major vehicle in which it is embodied, and their respective roles are closely interactive. There is no doubt of its importance in modern economic growth, or the contrast between its role in capitalist and pre-capitalist development. A major driving force of modern economies is the strong propensity to risk capital on new techniques that hold promise of improved profits, in strong contrast to the defensive wariness of the pre-capitalist approach to technology.

Since the eighteenth-century textile revolution, technical progress has been significant and continuous. Many of the products consumed today had no counterpart in the eighteenth century, and most production processes have been completely transformed. Furthermore, these economies are now geared to technical change in an organized and systematic fashion.

A good deal of research and development is now done in laboratories of large firms and by government departments. This trend started in 1876, when the Bell Telephone Company set up the first industrial research laboratory in the USA. Scientific and other research with ultimate economic applicability is carried out on a much larger scale now in universities than it was a century ago. The progress of technology is influenced to an important

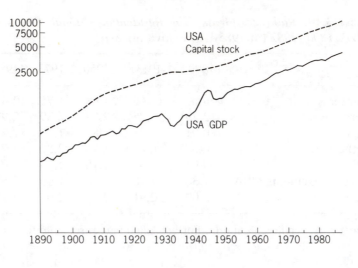

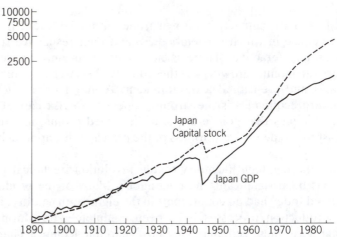

Graph 3.3 *Growth of Non-Residential Gross Capital Stock and GDP in the USA and Japan, 1890–1987*
Vertical scale is logarithmic and shows levels of GDP and gross capital stock in million 1985 US$. Horizontal scale is chronological and indicates decade intervals. The data are annual.

TABLE 3.11. *Gross Residential Capital Stock per Head of Population, 1950–1987* ($ at 1985 US relative prices)

	1950	1973	1987
Germany	4,935	12,270	19,032
Japan	1,598	4,858	8,740
UK	4,509	7,864	11,071
USA	12,314	18,626	22,807

Source: See Table 3.12.

degree by the amount of investment that is carried out, because this involves improvement engineering and learning by doing, which are always necessary in the practical implementation of new techniques. As investment activity has increased enormously over the past two centuries, there is yet another reason to expect that technical progress is built into the economic system. The role of individual inventors who develop their bright ideas outside an institutional framework has not disappeared, though their share of the action has declined markedly.[9] In 1901 82 per cent of US patents were issued to individuals, but by 1986 the percentage had fallen to 17 per cent.[10] The institutionalization of innovation made the pace of advance of knowledge less erratic, and probably helped speed up the rate of growth of technical potential.

A rough proxy measure of the pace of technical progress is performance of the lead countries as described in Chapter 2 above. In 1820–90, when the UK was the leader, its labour productivity grew at 1.2 per cent a year, whereas US performance has been appreciably better — averaging 2.2 per cent a year from 1890 to 1989. Although in most decades the US pace of advance did not deviate too far from average, this has not been invariably the case. In 1929–38, when demand was depressed, productivity growth fell well below the long-term trend, and in 1938–50, when demand expanded very rapidly, it boomed at 3.2 per cent a year for twelve years (Table 3.13). 1929–38 was a period of unprecedented economic collapse and 1938–50 included the wartime upsurge in output. If one looks at 1929–50 as a whole, the average growth in labour and total factor productivity was virtually the same as in 1913–29 and 1950–73. However, since 1973 both

TABLE 3.12. *Ratio of Gross Residential Capital Stock to GDP, 1950–1987* (at 1985 US relative prices)

	1950	1973	1987
Germany	1.48	1.21	1.44
Japan	1.02	0.53	0.64
UK	0.80	0.78	0.87
USA	1.43	1.32	1.32

Sources: For Germany, the UK, and USA the official estimates of residential capital were used (supplied by the Federal Statistical Office, the Central Statistical Office, and the Bureau of Economic Analysis respectively). German estimates in DM at 1980 prices were converted to 1985 prices by the implicit deflator in the national accounts (1.15837) and converted into 1985 dollars using the Eurostat Paasche PPP for residential construction of 3.21959 DM/$. UK figures in 1985 prices were converted to 1985 dollars by the Eurostat PPP of 0.82321£/$. US estimates in 1982 dollars were converted to 1985 dollars with the implicit national accounts deflator of 1.08399. The German authorities assume a 77-year life for residential capital; the US an 80-year life for 1–4 unit structures, 65 for bigger units, and 16 for mobile homes. UK housing stock is estimated by capitalizing the rateable values shown in the tax statistics, so there is no explicit length of life assumption. For Japan, no official estimates are available, so I used the stock estimates at 1965 prices given in K. Ohkawa and M. Shinohara, *Patterns of Japanese Economic Development: A Quantitative Appraisal*, Yale, 1979, p. 369 up to end 1969 updated to mid-1987 using national accounts information on residential investment, with scrapping derived from the historical estimates of earlier investment in Table A.39 of Ohkawa and Shinohara, and allowance for war damage. The Japanese estimates assume a 50-year life for housing, see ibid., p. 191. The bench-mark estimate at 1965 prices was converted to 1985 yen by the implicit national accounts deflator for residential investment (3.250934) and then converted into 1985 dollars using the Eurostat Paasche PPP for residential investment of 391.664 yen/$. For France and the Netherlands few value estimates are available and the estimates of residential capital cited below in Table 5.9 are based on physical indicators of the stock of rooms and houses, as described in A. Maddison, *Journal of Economic Literature*, June 1987, Table A.18.

TABLE 3.13. *Variations in Productivity Growth in the USA, 1890–1989 (annual average compound growth rate)*

	GDP per man-hour	Joint factor productivity		GDP per man-hour	Joint factor productivity
1890–1913	2.2	1.1	1950–73	2.5	1.8
1913–29	2.4	1.7	1973–89	1.1	0.6
1929–50	2.4	1.9			
1929–38	1.4	0.3			
1938–50	3.2	3.2			

Source: See Table 2.2.

labour productivity and total factor productivity have slowed down dramatically, and it is difficult to explain this.

It seems certain that it does not reflect error in the measurement of growth.[11] As it has affected all sectors of the economy (see Table 5.13), it does give genuine grounds for concern, both for the USA and for the follower countries, whose prospects are affected by the pace of advance at the technical frontier.[12]

For individual industries there is a logistic history with initial slow development, rapid acceleration, and then retardation of growth. This process was analysed in some detail by Kuznets and Burns,[13] and interpreted in large part as being due to what might now be called the 'product cycle' of technical progress,[14] where the S-shaped trend reflects changing elasticity of demand, as well as fading possibilities for innovation.

The evidence of ultimately diminishing returns in these individual industry and product histories has given rise, from time to time, to pessimistic assessments about future trends for the economy as a whole. Thus Kuznets wrote in 1941:

In the industrialized countries of the world, the cumulative effect of technical progress in a number of important industries has brought about a situation where further progress of similar scope cannot be reasonably expected. The industries that have matured technologically account for a progressively increasing ratio of the total production of the economy. Their maturity does not imply a complete cessation of further technological improvements, but it does imply that economic effects of further improvements will necessarily be more limited than in the past.[15]

This conclusion of Kuznets turned out to be premature, because the USA found other industries to take over the previous dynamic role of waning sectors, and productivity growth in agriculture accelerated considerably.

It has certainly been true historically that technical progress has impacted unevenly in different parts of the economy. There has been a long-term tendency towards slower productivity growth in the service sector of the economy than in commodity production. This may well mean that technical advance faces greater problems in this sector than elsewhere, and if this continues it will certainly dampen overall productivity growth in the future because employment in services is now so high and demand for them is rising fast.

STRUCTURAL CHANGE

In the past two centuries there have been large changes in the structure of expenditure, output, and employment. In 1820, food and clothing represented two-thirds of private consumption. Over time, increases in consumption have been concentrated on new products and items which used to be luxuries. By the late 1980s food and clothing represented only a quarter of private consumption and there was a decline in the share of private consumption in GDP. In 1820, it was probably about 85 per cent, but in 1988 it averaged only 58 per cent of GDP in our sixteen countries. By 1988, the share of government had risen to 18 per cent, and investment to 24 per cent.

The changes in demand patterns had a major impact in changing the structure of output. The changes in output affected the pattern of employment. But the pattern of employment was also influenced by differences in the rate and direction of technical progress. Over the long haul, this has been slowest in the service sector, and fastest in industry. In the post-war era, however, agriculture has shown the fastest productivity growth, and this acceleration together with the relative decline in demand for

Table 3.14. *Percentage Structure of Employment, 1870–1987* (average for 16 countries)

	Agriculture	Industry	Services
1870	48.8	26.9	24.3
1950	24.7	36.6	38.7
1960	17.5	38.7	43.8
1973	9.3	37.3	53.4
1987	6.0	30.5	63.5

Note: Agriculture includes forestry and fisheries; industry includes mining, manufacturing, utilities, and construction; services include all other activities including government (civilian and military).

Source: Appendix Table C.5 and A. Maddison, 'Economic Growth and Structural Change in the Advanced Countries', in I. Leveson and J. W. Wheeler, *Western Economies in Transition*, Croom Helm, London, 1980.

agricultural products has had a drastic effect on employment patterns.

Table 3.14 summarizes the change in employment structure since 1870. The detailed country figures are contained in Appendix Table C.5.

There has been a massive reduction in the share of employment in agriculture and a large rise in services. These two phenomena have operated more or less continuously in the same direction in all the countries. By contrast the industrial share has risen and fallen, peaking below 50 per cent of the employed population in all sixteen countries. In 1870 agriculture occupied half the population in these countries, in 1987 only 6 per cent. Service employment predominates, representing almost two-thirds of the total. Industry occupies 30 per cent — not much higher than in 1870.

INTERNATIONAL TRADE

Until the nineteenth century, prosperity gained through trade usually involved a considerable beggar-your-neighbour element because of the limited size of the world market and its rather slow growth. There was a whole succession of, for the times, prosperous countries whose fortunes were gained at the expense of ousted rivals. Thus, Spain and Portugal displaced Venice and Genoa as major traders with the East as the trade routes to Asia moved away from the Mediterranean. Dutch prosperity was achieved by eating into the Portuguese and Spanish trading empires and by impoverishing the southern Netherlands (present-day Belgium). British and French trade expansion in the eighteenth century was to a significant extent at the expense of the Dutch.

Since 1820, trade has greatly accelerated and has grown significantly faster than output — by 4 per cent a year compared with 2.7 per cent for GDP. This has improved resource allocation and productivity by better specialization and economies of scale, but the impact has varied over time.

From 1820 to 1913, trade was an important stimulus to growth. The international economy was reopened after the blockades of the Napoleonic Wars; the UK moved to a policy of free trade; the German states created a customs union (*Zollverein*). A series of important treaties achieved substantial reduction of tariff barriers.

TABLE 3.15. *Volume of Exports, 1720–1989* (annual average compound growth rates)

	1720–1820	1820–70	1870–1913	1913–50	1950–73	1973–89	1870–1989
Australia			4.8	1.3	5.8	4.5	3.8
Austria		4.7	3.5	−3.0	10.8	6.1	3.7
Belgium		5.4[b]	4.2	0.3	9.4	4.4	3.9
Canada			4.1	3.1	7.0	4.8	4.4
Denmark		1.9[c]	3.3	2.4	6.9	4.7	4.0
Finland			3.9	1.9	7.2	3.4	3.8
France	1.0[a]	4.0	2.8	1.1	8.2	4.6	3.5
Germany		4.8[d]	4.1	−2.8	12.4	4.7	3.5
Italy		3.4	2.2	0.6	11.7	4.9	3.8
Japan			8.5	2.0	15.4	6.8	7.5
Netherlands	−0.2		2.3[e]	1.5	10.3	3.6	3.7
Norway			3.2	2.7	7.3	6.7	4.3
Sweden			3.1	2.8	7.0	3.1	3.7
Switzerland		4.1	3.9	0.3	8.1	3.8	3.6
UK	2.0	4.9	2.8	0.0	3.9	3.9	2.3
USA		4.7	4.9	2.2	6.3	4.7	4.3
Arithmetic average		4.2	3.9	1.0	8.6	4.7	4.0

[a] 1715–1820. [b] 1831–70. [c] 1844–70. [d] 1840–70. [e] 1872–1913.

Sources: Appendix F, except for Netherlands, 1720–1820, which was derived from J. de Vries, *De economische achteruitgang der Republiek in de achttiende eeuw*, Amsterdam, 1959, p. 27. Figures are not adjusted for changes in geographic boundaries.

The mercantilist tradition of the eighteenth century was eliminated. The progress of technology favoured trade as freight costs
fell (through development of railways, steamships, and exploitation of the Suez Canal). The growth of trade relative to output was
a little slower from 1870 to 1913 than from 1820 to 1870, because
continental trade policies became more protectionist, particularly
for agricultural products; but the trade restrictions of those days
were mild by subsequent standards.

The 1913–50 period saw a relapse into neo-mercantilism, with
the blockades involved in two wars, the discriminatory policies,
higher tariffs, quantitative restrictions, exchange controls, and
other autarkic measures that were sparked off by the Great
Depression of 1929–32. As a result, trade grew at half the pace of
output from 1913 to 1950. This impeded efficiency and was a drag
on productivity growth.

After the Second World War, there was a succession of major
moves to restore liberal trade regimes. Quantitative restrictions on
non-agricultural products were dropped in the 1950s; successive
multilateral negotiations in the GATT reduced tariffs very substantially; the European Community abolished tariffs between its
members; and exchange controls virtually disappeared. These
moves contributed to efficient resource allocation in a direct way.
The establishment of institutional arrangements to guarantee
continuation of a liberal international economic order bolstered
confidence and was a major indirect stimulus to productivity
because it helped create a climate conducive to high rates of
investment.

Some of the growth-stimulating impact of trade faded after 1973
as the once-for-all effect of trade liberalization was absorbed, but
international trade has continued to rise faster than output and
pressures for a return to protectionism on a significant scale have
been resisted.

THE ROLE OF GOVERNMENT

In 1820, the franchise was limited to property owners, and the
state was mainly preoccupied with their interests. Expenditure was
concentrated largely on soldiery and police protecting property
and the national frontiers. Most of the population were poor and

TABLE 3.16. *Electorate as Percentage of Persons Aged 20 and Over*

	1869–73	1972–5		1869–73	1972–5
Austria	10.6	98.0	Norway	8.5	99.4
Belgium	3.7	94.3	Sweden	9.8	96.3
Denmark	25.8	98.5	Switzerland	38.7	83.5
France	42.0	87.5	UK	14.9	104.0
Germany	33.0	98.8	Average	17.8	96.2
Italy	3.5	99.4			
Netherlands	5.0	98.0			

Source: P. Flora and Associates, *State, Economy and Society in Western Europe 1815–1975*, vol. i, Macmillan, London, 1983, pp. 97–149.

TABLE 3.17. *Total Government Expenditure as a Percentage of GDP at Current Prices, 1913–1987*

	1913	1929	1938	1950	1973	1987
France	8.9	12.4	23.2	27.6	38.8	53.6[b]
Germany	17.7	30.6	42.4	30.4	42.0	47.3
Japan	14.2	18.8	30.3	19.8	22.9	33.9
Netherlands	8.2[a]	11.2	21.7	26.8	45.5	59.7
UK	13.3	23.8	28.8	34.2	41.5	45.2[b]
USA	8.0	10.0	19.8	21.4	31.1	37.0
Average	11.7	17.8	27.7	26.7	37.0	46.0

[a] 1910. [b] 1986.

Sources: France, 1913–38, numerator from L. Fontvieille, *Évolution et croissance de l'état français 1815–1969*, ISMEA, Paris, 1976, pp. 2124–9, and denominator from J. C. Toutain, *Le Produit intérieur brut de la France de 1789 à 1962*, ISMEA, Paris, 1987, pp. 155–7. United States, 1929–73, from *National Income and Product Accounts of the United States, 1929–1982*, Dept. of Commerce, Washington, DC, 1986, pp. 1–2, 43–4, 35–6. Otherwise from A. Maddison, 'Origins and Impact of the Welfare State, 1883–1983', *Banca Nazionale del Lavoro Quarterly Review*, Mar. 1984. Post-war years from OECD, *National Accounts*, various issues.

illiterate and the state did little to help or protect them. There were legal restraints on trade union activity and the official attitude to popular discontent was repressive.

As a result of increasing enfranchisement, political struggle, and the vast rise in real income, the nature of the state has been transformed. The major change was the emergence of 'welfare state' expenditures. From the 1880s onward there was a steady expansion in public provision for education and health, and over the past seventy years there has been a huge growth of pensions, sickness and unemployment benefits, and family allowances.[16] Pensions are the biggest of these payments. Their rise is due to increased benefit levels and to demographic factors. In 1870, people of 65 and over were only 5 per cent of the population. They are nearly 14 per cent now.

Table 3.18 shows the scale of these modern commitments, which are up to 40 per cent of GDP in continental Europe and lowest in the USA and Japan. By contrast, traditional commitments have changed much less since 1820.

The role of the state is more modest in terms of employment than in expenditure. Public employment ranges from about a fifth to a third of the total. Most involves provision of traditional services, education and health, but there is also a public enterprise sector. This is smallest in the USA, where it accounts for less than 2 per cent of total employment; it is largest in Austria, where it accounted for 12.4 per cent of total employment in the late 1970s.[17] In the 1980s the size of the public enterprise sector generally declined through privatization and now probably averages less than 5 per cent of total employment in these countries.

The impact of increased government activity has been to reduce poverty, to provide more equality of opportunity, and to reduce the uncertainty and insecurity connected with old age and sickness. It has also reduced the hardships of recession for the unemployed and provides some degree of built-in stability for the economy. The indexation of social transfers provides a hedge against inflation which private schemes could never supply. Government spending and taxation has only a mild direct effect on the vertical distribution of income[18] because the structure of taxes and levies is only mildly progressive, but state action in building up and equalizing the distribution of human capital and in promoting the economic security and buying power of labour has undoubtedly

TABLE 3.18. *Structure of Government Expenditure as a Percentage of GDP, 1981*

	France	Germany	Japan	Netherlands	UK (1978)	USA
Traditional commitments						
Military	3.8	2.9	0.9	3.3	5.0	4.7
Civil	3.7	3.9	3.3	6.1	3.6	3.7
Debt interest	2.2	2.2	3.6	5.4	5.0	2.7
Total	9.7	9.0	7.8	14.8	13.6	11.1
Modern commitments						
Economic Services	3.9	4.9	6.0	3.8[a]	4.3	3.2
Education	5.9	5.2	4.9	7.0	5.4	5.7
Health	6.4	6.8	4.7	6.6	5.0	3.7
Housing etc.	3.7	2.3	2.9	5.0	2.2	0.8
Pensions	11.9	12.6	4.7	12.9	7.4	6.7
Other Transfers	5.3	4.1	2.2	4.8	3.1	1.0
Total	37.1	35.9	25.4	40.1	27.4	21.1

[a] 1978.

Sources: A. Maddison, 'Origins and Impact of the Welfare State', *Banca Nazionale del Lavoro Quarterly Review*, Mar. 1984, OECD, *National Accounts 1975–87*, vol. ii, Paris, and OECD, *Economic Studies*, Spring 1985, p. 52.

had an impact in equalizing the primary distribution of income.[19]

Over the long run there seems little doubt that state action has strengthened the forces making for economic growth and stability. It has also made capitalist property relations and the operation of market forces more legitimate by removing most of the grievances which motivated proponents of a socialist alternative. The 'socialist' parties in these countries have now generally abandoned the aim of nationalizing industry or significantly interfering with the operation of market forces.

The increase in state spending has of course had costs in terms of efficiency. In some areas the generosity of social transfers reduces incentives and lowers labour force participation, the downward flexibility of prices has disappeared, government 'economic services' are often a euphemism for subsidies to inefficient enterprise, and there is some bureaucratic waste involved in the 'churning' process of high social levies and benefits. However, these costs are modest in comparison with those involved in Communist countries, where the level of state intervention has been very much higher.

COLONIALISM

Economic progress in the advanced capitalist countries stimulated growth in the rest of the world in so far as it provided opportunities for trade, investment, and transfer of technology. Since the 1960s, there has also been a modest amount of development aid. In spite of this the level of real income in the rest of the world is only a fraction of the Western level. This is partly due to indigenous institutional and social arrangements less favourable to growth, but it was also due in some degree to Western exploitation through various forms of colonialism from 1500 to the 1960s.

The monopolistic profits and privileged markets of merchant capitalism were certainly influential in establishing the early economic leadership of the Netherlands and then the UK. In the course of the nineteenth century, the monopolistic elements of merchant capitalism were generally abandoned in favour of free trade imperialism. The scramble for colonies, treaty ports, and spheres of influence accelerated and involved eleven of our sixteen countries to some degree. The benefits of this latter-day imperi-

TABLE 3.19. *Categories of Government Revenue as a Percentage of GDP at Current Market Prices, 1987*

	Total current revenue	Social security levies	Direct taxes	Indirect taxes	Other
France	49.4	19.1	9.5	14.7	6.0
Germany	44.4	16.2	12.2	12.2	3.6
Japan	33.2	8.7	13.0	8.3	3.2
Netherlands	53.2	20.6	14.1	11.9	6.7
UK	40.7	6.8	14.3	16.0	3.6
USA	31.9	7.2	14.0	8.2	2.6
Average	42.1	13.1	12.9	11.9	4.2

Source: OECD, *National Accounts 1976–88*, vol. ii, Paris, 1990.

alism were too marginal to figure significantly in the explanation of Western growth. In fact the overall influence was negative for the West because it was a source of conflict and war. Since colonialism was abandoned, the growth of Western economies has accelerated, as it has in most of the former colonies.[20]

NOTES ON CHAPTER 3

1. Earlier writers who stressed the existence of a general turning-point in 1820 are J. A. Schumpeter, A. Spiethoff, and M. von Tugan-Baranowsky. The latter was the first to date capitalist dynamics from the second quarter of the 19th century; see his *Studien zur Theorie und Geschichte der Handelskrisen in England*, Fischer, Jena, 1901, p. 41. Spiethoff endorsed this view in 1923: 'Jedenfalls scheint mir erst mit den 1820er Jahren der recht eigentlich kapitalistische Kreislauf der Wechsellagen zu beginnen', *Handwörterbuch der Staatswissenschaften*, Jena, 1923, p. 47; and Schumpeter also adopted it in his *Business Cycles*, vol. i, McGraw-Hill, New York, 1939, p. 254. Rostow differed from this view and suggested that there was a long series of staggered take-offs in these countries in the course of the 19th century. Gerschenkron also believed this, but our evidence rejects this hypothesis except in the case of Japan. See W. W. Rostow, *The Stages of Economic Growth*, Cambridge, 1962, and A. Gerschenkron, *Economic Backwardness in Historical Perspective*, Praeger, New York, 1965.
2. See Y. Hayami and V. W. Ruttan, *Agricultural Development: An International Perspective*, Johns Hopkins, Baltimore, 1971, for an analysis of the response of technology to different resource/population ratios in Japan and the USA. See OECD, *National Policies and Agricultural Trade*, Paris, 1987, for a detailed quantification of the value of agricultural subsidies.
3. See W. S. Jevons, *The Coal Question*, Macmillan, London, 1865.
4. D. H. Meadows *et al.*, *The Limits to Growth*, Universe, New York, 1972.
5. The development of opinion on hours is described in E. J. Hobsbawm, 'Custom, Wages and Work-Load in Nineteenth Century Industry', *Labouring Men*, Doubleday, New York, 1967, and the evolution in hours in M. A. Bienefeld, *Working Hours in British Industry: An Economic History*, Weidenfeld & Nicolson, London, 1972. For German working hours, see W. H. Schröder, 'Die Entwicklung der Arbeitszeit im sekundären Sektor in Deutschland 1871–1913', *Technikgeschichte*, no. 3, 1980, p. 267, who quotes annual figures for Nuremberg industry showing a rise in weekly hours from 61.1 in 1821 to a peak of 66.5 in 1870.
6. See R. C. O. Matthews, C. H. Feinstein, and J. C. Odling-Smee, *British Economic Growth 1856–1973*, Oxford, 1982, ch. 4 and Appendix E for educational stock estimates back to 1805 for males in England and Wales. Comparative enrolment figures back to 1830 can be found in R. A. Easterlin, 'Why isn't the Whole World Developed?', *Journal of Economic History*, Mar. 1981, pp. 1–19. I give comparative estimates for developed and developing countries in A. Maddison, *The World Economy in the Twentieth Century*, OECD, Paris, 1989, p. 78.
7. See T. W. Schultz, 'Investment in Human Capital', *American Economic Review*, Mar. 1961, pp. 1–17, for a seminal contribution. Our Table 3.8 gives only a very crude measure of the growth of human capital. It omits post-formal training and learning experience and the quality of education. Nevertheless, I feel that it probably gives a valid impression of the pace of improvement over time. Comparisons of human capital levels between these countries are a more delicate matter. Post-formal education is better developed in France, Germany, and Japan than in the UK and USA, and average cognitive

achievement in secondary education is higher in the five follower countries than in the USA. See OECD, *Learning Opportunities for Adults*, Paris, 1977, for a summary of post-formal education facilities in nine countries; for cognitive achievement, see T. N. Postlethwaite and D. E. Wiley (eds.), *Science Achievement in 23 Countries*, Pergamon, Oxford, 1991.

8. In some of the theoretical literature on growth there has been a tendency to assume long-run stability of capital output ratios: see N. Kaldor, 'Capital Accumulation and Economic Growth', in F. A. Lutz and D. C. Hague (eds.), *The Theory of Capital*, Macmillan, London, 1961, pp. 177–222, who took this proposition to be a 'stylised fact'. Gallman's estimates in S. L. Engerman and R. E. Gallman, *Long Term Factors in American Economic Growth*, Chicago, 1986, p. 186, suggest that the rise in the US capital/output observable on the left of Graph 3.3, was the continuation of a process going back as far as 1840.

9. The primordial role of large corporations was asserted by J. A. Schumpeter, *Business Cycles*, McGraw-Hill, New York, 1939, p. 1044: 'economic progress in this country is largely the result of work done within a number of concerns at no time much greater than 300 or 400'. The opposite view is represented by J. Jewkes, D. Sawers, and R. Stillerman, *The Sources of Invention*, St. Martin's Press, New York, 1958, who found that over half of 61 significant 20th-century inventions were produced by individual rather than company research.

10. See *Historical Statistics of the United States, Colonial Times to 1970*, US Dept. of Commerce, pt. 2, Washington, DC, 1975, p. 958, and *Statistical Abstract of the US 1988*, Bureau of the Census, Washington, DC, p. 518.

11. See E. F. Denison, *Trends in American Economic Growth 1929–1982*, Brookings Institution, Washington, DC, 1985.

12. See W. J. Baumol, S. A. B. Blackman, and E. N. Wolff, *Productivity and American Leadership*, MIT, 1989, for a more insouciant reading of the evidence.

13. See S. Kuznets, *Secular Movements in Production and Prices*, Houghton Mifflin, Boston, 1930, and A. F. Burns, *Production Trends in the United States since 1870*, NBER, New York, 1934.

14. See OECD, *Gaps in Technology: Analytical Report*, Paris, 1970, pp. 222–5, for an illustration of the point and references to the literature.

15. See S. Kuznets, *Economic Change*, Norton, New York, 1953, p. 281. He made the same point earlier in *Secular Movements in Production and Prices* (above, n. 13), p. 11, where he quotes Julius Wolf's 'laws of retardation of progress'.

16. See A. Maddison, 'Origins and Impact of the Welfare State, 1883– 1983', *Banca Nazionale del Lavoro Quarterly Review*, Mar. 1984, pp. 55–87, for a more detailed analysis of the reasons for the growth of state activity.

17. See L. Pathirane and D. W. Blades, 'Defining and Measuring the Public Sector: Some International Comparisons', *Review of Income and Wealth*, Sep. 1982, pp. 261–89.

18. See M. Sawyer, 'Income Distribution in OECD Countries', *OECD Economic Outlook, Occasional Studies*, July 1976, pp. 3–36.

19. See S. Kuznets, 'Economic Growth and Income Inequality', *American Economic Review*, Mar. 1955, who suggests that capitalist development at first involved increased inequality, and then took a U-turn towards increased equality. For an extensive review of the literature on the history of income distribution and of causal influences, see the contribution of F. Kraus to P. Flora and A. J. Heidenheimer (eds.), *The Development of Welfare States in Europe and America*, Transaction Books, London, 1981.

20. For a more detailed quantification of the effects of colonialism, see A. Maddison, *Class Structure and Economic Growth: India and Pakistan since the*

Moghuls, Allen & Unwin, London, 1971; A. Maddison and G. Prince (eds.), *Economic Growth in Indonesia, 1820–1940*, Foris, Dordrecht, 1989; and A. Maddison, 'The Colonial Burden: A Comparative Perspective', in M. Scott and D. Lal (eds.), *Public Policy and Economic Development*, Oxford, 1990, pp. 361–75.

4

FLUCTUATIONS IN THE MOMENTUM
OF GROWTH

It is clear from the preceding analysis that the process of capitalist development has not been smooth. There have been distinct and important phases of development which are worthy of study, definition, and causal interpretation. I distinguish four phases, which I shall describe later, covering periods of unequal length: 1820–1913, 1913–50, 1950–73, and 1973 onwards. There have also been shorter-term fluctuations, usually called business cycles. My primary interest is not in these, but in major changes in trend which are distinguished from each other by changes in the institutional-policy mix and usually initiated by some sort of 'system shock' which upsets established patterns of international intercourse.

Before presenting my own diagnosis, it is useful to trace the history of cyclical or wave analysis, because my quantitative empirical approach is not the only one available. In the past there have been a number of theories concerning the nature of long waves in economic activity. These were revived and augmented in the 1970s after a period when even the business cycle was considered obsolete and the long-wave hypothesis was regarded as quaint.[1]

The unfortunate thing about revivalist approaches to new problems is that the adherents are often single-minded enthusiasts, so that the analytical apparatus of the old theories is rehabilitated *in toto* in spite of remediable weaknesses.

CYCLE ANALYSIS

Cyclical analysis for the capitalist period started with Clement Juglar in 1856. He emphasized periodicity in economic activity whereas most earlier writers had tended to interpret interruptions to growth as random financial crises. Juglar also believed that cycles were roughly synchronous in France, the UK, and USA.[2] In his major work on cycles his attention was mainly concentrated on monetary phenomena — expansions or contractions in central bank activity, rates of interest, prices of key commodities, etc., plus narrative 'business annal' material. Although it is frequently asserted that Juglar found cycles of a characteristic length of nine years, this is not in fact true. His cycles for France average seven years with a range from three to eighteen years, and for the UK six years with a range from two to ten years.

For several decades the quantitative indicators available to cyclical analysts were similar to those used by Juglar, though they were later augmented to include price indices, and data on output and foreign trade. A more sophisticated causal analysis was also developed, such as one finds in the study by the Russian economist Tugan-Baranowsky on the nineteenth-century cycle in the UK.[3]

The ultimate refinement in statistical analysis of business cycles was the massive effort of the National Bureau of Economic Research (NBER) in the USA. The first phase was a comprehensive collection of narrative data stretching back to the beginning of the nineteenth century with a cyclical periodization for seventeen countries. The second phase was publication of a series of reference' cycles for four countries (France, Germany, Great Britain, and the USA) based mainly on monthly quantitative data, which start in 1854 for the last two countries, in 1865 for France, and in 1879 for Germany.[4] The number of monthly series for the USA was nineteen for 1860 rising to 811 in 1942 (plus 161 annual indicators). The NBER established its reference cycles by plotting most of this information in de-seasonalized form, and by iterative procedures of inspection, derived a cluster of roughly concurrent fluctuations. Thus its central concept of economic activity was a somewhat fuzzy cocktail rather than a clearly defined measure of aggregate economic activity. Its main use was as a sensitive warning indicator of turning-points in business activity, with indicators classified as leading, coincident, or lagging. The reference cycle has become part of the official statistical armoury of the USA for forecasting purposes, though it is of course supplemented by the more articulate short-term models on which other countries place main reliance. For the period 1857 to 1978 the NBER established twenty-eight successive peak-to-trough movements for the United States, giving a recession on average every four years, with a variation from two-and-a-half to nine-and-a-half years. For other countries the average duration was found to be longer: fifty-three months for France, sixty-two for the UK, and sixty-four for Germany for pre-war years. The NBER cycles are not adjusted to eliminate trend, so they are not measures of oscillation in economic activity, and register recessions only when there is an absolute fall in the relevant indicators.[5] However, the NBER technique of using monthly and rather volatile series does pick up more cycles than would a GDP index based on annual data, and those reference cycles that do correspond with GDP movements do not always have exactly the same dates.[6] The NBER approach

TABLE 4.1. *Amplitude of Recessions in Aggregate Output, 1870–1989*

	Maximum peak–trough fall in GDP or lowest rise (annual data)			
	1870–1913	1920–38	1950–73	1973–89
Australia	−17.1	−12.8	+0.9	+0.2
Austria	−2.3	−22.4	+0.1	−0.4
Belgium	−0.2	−7.9	−0.8	−1.5
Canada	−7.7	−29.6	−0.7	−3.2
Denmark	−2.7	−2.9	−0.7	−1.6
Finland	−4.2	−4.0	+0.5	+0.1
France	−6.5	−14.7	+2.5	−0.3
Germany	−3.2	−16.9	−0.1	−1.6
Italy	−6.7	−10.1	+1.6	−2.6
Japan	−7.4[a]	−7.3	+4.3	−1.2
Netherlands	n.a.	−9.5	−0.3	−2.1
Norway	−3.0	−8.3	−0.9	+0.3
Sweden	−5.5	−6.2	−0.2	−1.6
Switzerland	n.a.	−8.0	−2.1	−8.6
UK	−4.1	−8.1	−0.2	−3.5
USA	−8.2	−29.6	−1.4	−2.6
Arithmetic average	−5.6	−12.4	+0.2	−1.8

[a] 1885–1913.

Source: Appendix A.

is a useful tool in interpreting quantitative economic history, but a major problem is that it yields no satisfactory measure of the amplitude of fluctuations because of the difficulty of producing a meaningful summary measure from such heterogeneous data. Thus one cannot use the reference cycle itself to distinguish major and minor cycles, in the same way that one can with simpler measures of industrial output or GDP fluctuations.

Hence my preference is for rather simple measures of annual movements in aggregate activity, which reveal clearly the big changes in the severity of recessions that have appeared systematically across our sixteen countries in the past century, as illustrated in Table 4.1. This table shows that peacetime business cycle

TABLE 4.2. *Amplitude of Recessions in Exports, 1870–1989*

	Maximum peak–trough fall or smallest annual rise in export volume (annual data)			
	1870–1913	1920–38	1950–73	1973–89
Australia	−32.2	−19.7	−7.6	−3.6
Austria	n.a.	−48.7	−7.3	−5.5
Belgium	−13.1	−31.8	−9.6	−6.8
Canada	−13.9	−40.6	−6.1	−10.3
Denmark	−25.0	−20.3	−6.7	−4.1
Finland	−20.9	−15.7	−13.4	−17.4
France	−12.9	−47.3	−12.0	−4.0
Germany	−14.2	−50.1	+2.3	−11.5
Italy	−30.6	−69.1	−9.1	−8.5
Japan	−23.7	−18.9	0.0	−2.3
Netherlands	n.a.	−33.4	+2.6	−3.8
Norway	−7.7	−16.0	−7.1	−3.3
Sweden	−11.0	−37.0	−10.7	−11.6
Switzerland	n.a.	−50.2	−4.2	−8.2
UK	−12.5	−37.7	−8.0	−2.2
USA	−18.9	−47.8	−14.3	−18.8
Arithmetic average	−18.2	−36.5	−7.0	−7.6

Source: Appendix F.

history has been much milder since the Second World War than before, and that the 1920–38 period was generally much worse than 1870–1913. Except in 1929–33, when the Depression hit every country, the weighted average of cyclical movements for the sixteen countries as a group is dampened by the fact that individual country cycles are not synchronized. Before 1870 data on annual changes in GDP are not available for most countries, but it would seem that average cyclical experience was not too different from that of 1870–1913.[7] Table 4.2 shows the cyclical record for foreign trade. It confirms the pattern shown by GDP movements, with notably smaller cycles since the Second World War. Graphs 4.1 and 4.2 are intended to show both cyclical volatility and differences in growth trends in different phases or capitalist development. The striking thing in both graphs is the great volatility of the 1913–50 period, and the markedly faster growth since 1950.

LONG-WAVE ANALYSIS

Although cyclical analysts had made distinctions between big and small recessions, and there had been some discussion of the Great Depression (in prices) in the last quarter of the nineteenth century, it is significant that the idea of recurrent long waves in capitalist development did not emerge until the First World War, i.e. about fifty years later than cycle analysis, and only after the rhythm of development had been very dramatically broken.

The main figures in long-wave analysis are N. D. Kondratieff, S. Kuznets, and J. A. Schumpeter. All of them drew heavily on cyclical-type indicators to test their ideas quantitatively.

Kondratieff

Kondratieff was a Russian economist, whose work on long waves was done in the 1920s as director of the Business Cycle Research Institute in Moscow. He distinguished three kinds of cycles: long ones of fifty years' duration, middle ones of seven to ten years', and short ones of three to four years'. He measured long waves by a double decomposition of time series — eliminating the trend and showing the deviations from it smoothed with a nine-year moving average. The nine-year average was enough to remove the

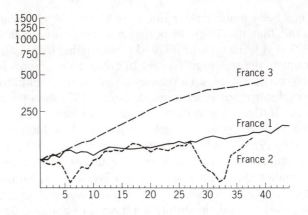

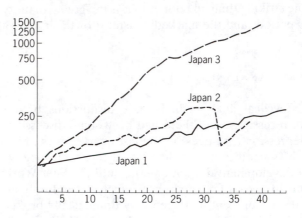

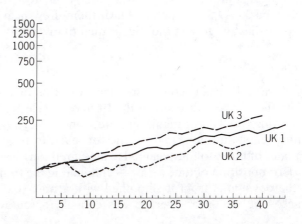

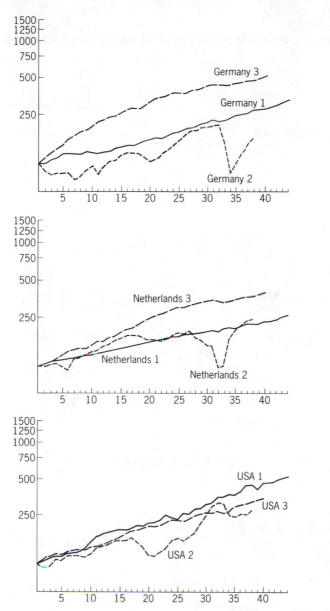

Graph 4.1 *A Comparison of Volume Movement of GDP in Different Periods, 1870–1989*

Vertical scale is logarithmic and shows growth performance; horizontal scale superimposes three periods: (1) 1870–1913; (2) 1913–50; and (3) 1950–89. Initial year of each period = 100. Quinquennial intervals are marked on the horizontal scale.

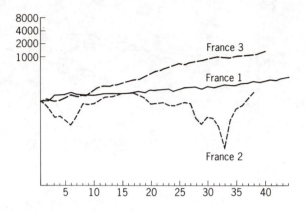

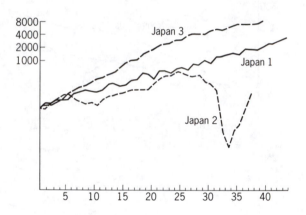

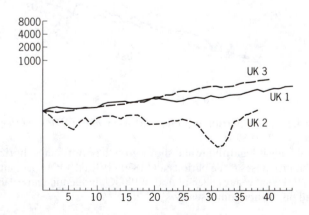

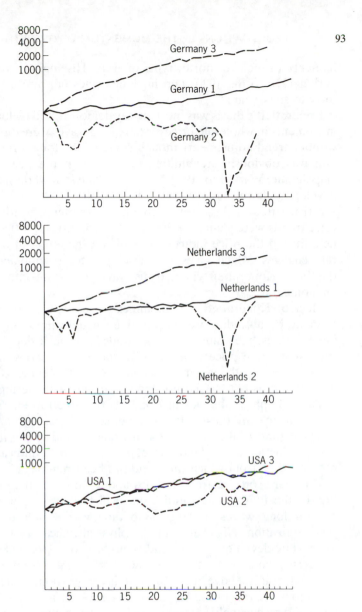

Graph 4.2 *A Comparison of Volume Movement of Commodity Exports in Different Periods, 1870–1989*
Vertical scale is logarithmic and shows level of exports; horizontal scale is chronological and superimposes three periods: (1) 1870–1913; (2) 1913–50; and (3) 1950–89. Initial year of each period = 100.
Quinquennial intervals are marked on the horizontal scale.

influence of the two shorter types of cycle. His analysis covers the period 1770 to the 1920s and his long cycles fall into a range of forty to sixty years.[8]

Kondratieff's thesis was most clearly demonstrated by long-term movements in wholesale prices, where long waves were discernible without trend adjustment, though some of the long-term oscillation was obviously attributable to wars (e.g. the peaks in the Napoleonic Wars and 1914–20). He analysed wholesale price developments for France, the UK, and the USA, and it is not surprising that in these relatively open economies he found that price trends were similar in the different countries, particularly as he adjusted the price indices to eliminate the effect of exchange rate changes, which gives the series greater resemblance.[9] On this basis Kondratieff claimed his waves to be an international phenomenon.

Most of Kondratieff's other indicators contain a strong price element, because they are expressed in current values: e.g. wages, interest rates, the value of foreign trade, and bank deposits. Not surprisingly, the price component of these value series moves in the same way as the general price indices, so this evidence for his wave theory is not in fact independent of his first offering.

The only physical series in Kondratieff's repertoire in his most famous article are those relating to per capita[10] coal production in England (and coal consumption in France), and to pig iron and lead production in England. Here, as with his value indicators, he presents data from which the trend has been removed.

There are some distinct oddities about Kondratieff's presentation of the four physical indicators, which at first sight seem to contain long waves of large amplitude with a fair degree of synchronization. His charts for the physical indicators are shown as absolute deviations from trend. Thus he shows UK coal production 186 points above trend in 1869, 245 points below in 1894, and 164 points above in 1910. But in proportionate terms the deviations are much smaller: 5.4 per cent, −5.1 per cent, and 2.8 per cent respectively. He also follows the highly questionable practice of juxtaposing two series on the same graph to suggest that the amplitude of their movement is similar, this effect being secured by using quite different scales for each. From this graph it appears that UK coal output is more volatile than French coal consumption, whereas the proportionate swings in France were

TABLE 4.3. *Kondratieff's Long-Wave Chronology*

	Rise	Decline
1. First long wave	1780s–90s to 1810–17	1810–17 to 1844–51
2. Second long wave	1844–51 to 1870–5	1870–5 to 1890–6
3. Third long wave	1890–6 to 1914–20	1914–20 to ?

bigger than in the UK. Worse problems arise in his graph for British pig iron and lead production, because there is the further complication that he there compares two series with totally different trends. Pig iron output rose about fourfold over the period he covered, and lead output fell to less than a tenth of its original level.[11]

Kondratieff concluded tentatively that there had been three long waves in economic 'life' (a rather vague term, but one that is clearly intended to include output as well as price movements). His chronology refers not to particular years but to spans, and he distinguishes only two phases, the rise and fall, in each wave. He does not discuss the amplitudes of these waves, which vary between series, but they are clearly considered large enough to exclude the need for discussion of growth trends. His dating is as in Table 4.3.

There are several problems with Kondratieff's approach. The first is his failure to establish that long waves exist as more than a monetary phenomenon. He fails to show the existence of broad movements in the volume of output that even remotely correspond to our present measures of aggregate economic activity. The second problem is that the trend is taken out and discarded as if it were irrelevant to the discussion. Thus, in comparing UK and US growth between 1820 and 1989, one finds British GDP has risen about twenty-seven-fold, and American by more than 450-fold. This fact is left out when the time series are decomposed for wave analysis, but such very different trends transform the nature and operational significance of any long waves that may be discerned. The third problem is that double decomposition of time series to eliminate trend and smooth out cycles blurs the impact of major historical events. Thus, Kondratieff's chronology pays no attention to the impact of the First World War, and later long-wave analysts tend to brush off the catastrophic 1929–33 recession and the Second World War as well. Finally, Kondratieff fails to offset these empirical shortcomings by giving plausible causal explanations as to why capitalist development should involve long waves as a systematic phenomenon. In the USSR this problem involved Kondratieff in ideological difficulties because his wave theory seemed to conflict with the more fundamental Marxist expectation of the ultimate breakdown of capitalism.[12]

There is no doubt that Kondratieff's contribution to long-wave analysis was fundamental in spite of its weakness,[13] because he fully adumbrates the three-cycle schema later developed by Schumpeter, and his statistical technique was the same that Kuznets later used to distinguish 'secondary secular movements'. Furthermore, he pointed to the likelihood of poor terms of trade for agriculture in periods of decelerated development — a point given major stress later by Arthur Lewis.

Kuznets and Abramovitz

Chronologically, the next development in long-wave analysis was Kuznets's work on 'secondary secular movements', published in 1930.[14] Kuznets's basic technique for identifying long waves was the same as Kondratieff's, i.e. to look at smoothed detrended series, though Kuznets made a special point of not eliminating population movements. His investigation was more detailed, involving careful analysis of fifty-nine series, most of which represented both physical output and the relevant price variance for particular commodities.[15] Kuznets did not claim that these indicators could be added to provide a meaningful picture of aggregate economic activity, and he did not use aggregative indicators for sectors such as agricultural or industrial production, which were available when he wrote.

His major conclusions are: (1) that 'secondary secular variations in production are in most cases similar to those in prices, the latter following a rather general course in agreement with the well-known historical periods of the rise and fall in the general price level' (p. 197); (2) he found a much shorter periodicity than Kondratieff, 'about 22 years as the duration of a complete swing for production and 23 years for prices' (p. 206); (3) most fundamentally, he did not think there was enough evidence to conclude that these secondary secular variations were major cycles. They were 'rather specific, historical occurrences' (p. 258). There is 'an absence of factors that would explain the periodicity' (p. 264).

Kuznets did not attempt to cluster his individual series to present a global chronology of long waves in economic life, nor did he analyse the synchronization of the series.[16] However, it is clear from other evidence that in the period Kuznets covered there were rather large depressions in the USA at intervals of fifteen to twenty years. This is directly observable in indices of industrial production (including construction), which Arthur Lewis has prepared (see Table 4.4). It is also clear that the recession/depression sequence was different in France, Germany, and the UK, which is the major reason why the aggregate performance of these four countries (which Lewis calls 'the core', to distinguish them from 'the periphery' — the rest of the world) is more stable than they are individually. In comparing the cyclical performance of these countries, it is useful to keep in mind the differences in their long-run growth performance. A country like France or the UK, with slow growth, is likely to have more small recessions than Germany or the USA, which had much higher growth. A rough measure of how far recessions fell below the potential growth path is to combine the trend and the cyclical amplitude: e.g. the average French recession involved a fall of 6.7 per cent from trend (4.1 + 2.6 per cent), and the average German recession a fall of 7.5 per cent from trend (3.2 + 4.3 per cent).

After his early study of secondary secular movements, Kuznets moved on to fundamental definitional work on the rationale (scope, valuation, and net-ness) for GDP as an aggregate economic indicator within a system of national accounts, and produced historical estimates of US economic development which made it possible to analyse long-term movements in economic life on a much more satisfactory conceptual basis than the cocktail approach that virtually all economic analysts had previously been forced to use. Furthermore, Kuznets successfully stimulated and inspired replication of his work by scholars in many other countries. This accounting approach has revolutionized the study of growth and greatly facilitates testing of long-wave analysis.

From time to time after 1930 Kuznets returned to long-swing analysis in a rather tentative way. Unlike his disciples, he himself

TABLE 4.4. *Amplitude and Duration of Cycles in Industrial Production (including Construction), 1870–1913*

	Peak year	Trough year	% amplitude of peak–trough movement	Duration of recession (years below peak)
France	1878	1879	−0.7	1
	1882	1885	−10.6	6
	1892	1893	−3.4	1
	1894	1895	−4.4	1
	1899	1902	−7.9	5
	1907	1908	−1.4	1
	1909	1910	−2.5	1
	1912	1913	−2.1	1
	Average amplitude		−4.1	
	Percentage of years below peak			39.5
	Trend growth rate		2.6	

			Amplitude	Years
Germany	1873	1874	−0.5	1
	1876	1877	−5.7	2
	Average amplitude		−3.2	
	Percentage of years below peak			7.0
	Trend growth rate		4.3	
UK	1876	1879	−4.1	3
	1883	1886	−9.7	4
	1891	1893	−6.3	3
	1902	1903	−2.1	2
	1907	1908	−8.0	4
	Average amplitude		−6.0	
	Percentage of years below peak			37.2
	Trend growth rate		2.1	
USA	1872	1876	−14.8	6
	1883	1885	−6.0	2
	1890	1891	−1.5	1
	1892	1894	−15.9	2
	1895	1896	−7.1	1
	1903	1904	−6.3	1
	1906	1908	−16.7	2
	1910	1911	−4.2	1
	Average amplitude		−9.1	
	Percentage of years below peak			37.2
	Trend growth rate		4.7	

TABLE 4.4. (Cont.) *Amplitude and Duration of Cycles in Industrial Production (including Construction), 1870–1913*

	Peak year	Through year	% amplitude of peak – trough movement	Duration of recession (years below peak)
Four countries combined = core	1873	1874	−1.2	2
	1876	1877	−0.4	1
	1883	1885	−5.0	2
	1892	1893	−5.2	2
	1899	1900	−0.3	1
	1903	1904	−2.1	1
	1907	1908	−9.5	1
Average amplitude			−3.4	
Percentage of years below peak				23.3
Trend growth rate			3.6	

never called them 'cycles', as the word implies greater certainty about such phenomena and their periodicity than Kuznets concedes. In 1956 he did advance a tentative chronology of long swings in GDP for eight countries,[17] but the periodization looks very odd, because the logic of the analysis calls for a declining phase in the decades following 1946, and Kuznets later dropped this one attempt to suggest a general chronology for long swings.

Kuznets's work on long swings was only a small part of his output and was concentrated on US experience. His most affirmative article was a 1958 study on population growth,[18] which found the 'long-swing' hypothesis most plausible in relation to US population growth and to 'population-sensitive' components of capital formation such as housing and railway construction. It was applicable in weaker and sometimes inverse form in other national accounting aggregates.

Kuznets had several disciples in long-swing analysis whose work he generally endorsed. Abramovitz is the most interesting of these,[19] because he has made the most ambitious attempt to discern long swings in aggregate US economic activity and has veered between more positive affirmation of long swings than Kuznets and outright recantation, in the sense that he did not find valid evidence for the phenomenon in the post-war period. He also embraces a wider range of demand and supply considerations than Kuznets, who concentrated on the supply side.[20]

Abramovitz distinguishes waves of acceleration and retardation in US growth with an average duration for the full swing of fourteen years and a variance from six to twenty-one years, using NBER reference cycle indicators back to the 1820s. He uses a cocktail of twenty-nine indicators including GNP. He smooths his series by a rather complicated procedure, designed to eliminate NBER reference cycles, before removing the trend. He found that the turning-points of his different series 'cluster in relatively narrow bands of years'. He therefore produced a general chronology with nine swings between 1814 and 1939.

Even at his most affirmative, Abramovitz was basically cautious about the nature of long swings. Thus in 1959 he wrote: 'It is not yet known whether they are the result of some stable mechanism inherent in the structure of the US economy, or whether they are set in motion by the episodic occurrence of wars, financial panics, or other unsystematic disturbances.' By 1968 he concluded that

Kuznets cycles were 'a form of growth which belonged to a particular period in history' (1840–1914), which had disappeared thereafter.

Schumpeter

The most complex cycle system was propounded by Schumpeter. He had a basic Kondratieff long wave of fifty years, on each of which he superimposed six eight- to nine-year 'Juglars', each in turn being crowned by three forty-month 'Kitchin' cycles.[21] Schumpeter insisted on the empirical regularity of his schema as if the basic facts about these three cycles had been well established, whereas there are great doubts about all three, as well as the legitimacy of his nomenclature. Kitchin's paltry contribution to the literature is lean meat indeed compared with that of the NBER, and Juglar never claimed to have demonstrated the existence of an eight- to nine-year rhythm. In fact, the NBER had already demonstrated rather wide variance in the length of cycles, so that there was little ground for distinguishing Juglars and Kitchins. Furthermore, Schumpeter distinguished only the length of his three types of cycle and said nothing about their amplitude.

Schumpeter's treatment of statistical material is illustrative rather than analytic and is at times rather cavalier. He uses business annal material of the type favoured by his former colleague Spiethoff, or by Tugan-Baranowsky, both of whom had an obvious influence on his views. He also uses an NBER type of statistical 'cocktail' material in pulse charts of industrial production, prices, interest rates, deposits, and currency circulation (p. 465). He makes passing reference to national income analysis (p. 561), but elsewhere refers to the concept of total output as a 'meaningless heap' (p. 484), national income as a 'highly inconvenient composite' (p. 561).[22]

Schumpeter's cycle analysis runs to 1,050 pages and is highly discursive. Judged on its statistical evidence alone, it would have been long discredited. Its power lies in the imaginative theory he supplies to explain long waves and the highly illuminating commentary on many aspects of German, British, and American economic history. He argues that each wave represented a major upsurge in innovation and entrepreneurial dynamism. Although writing in the later 1930s, he was remarkably sanguine about the long-run productive potential of capitalism. For him, depressions were a necessary part of the capitalist process. They were a period of creative destruction, during which old products, firms, and entrepreneurs were eliminated and new products were conceived. In fact, Schumpeter dismissed the 1929–33 US recession much too lightly. He says: 'The depression that ran its course from the last quarter of 1929 to the third quarter of 1932 does not prove that a secular break has occurred in the propelling mechanism of capitalist production because depressions of such severity have repeatedly occurred — roughly once in every fifty-five years.'[23] He then quotes the 1873–7 period as if it were a precedent for 1929–33. Such a comparison is totally misleading. In the earlier period the peak–trough fall in US industrial production was 14.8 per cent; in the later one, 44.7 per cent! There is no earlier parallel to the 1929–33 fall either in its amplitude or international incidence.

Like most long-wave analysts, Schumpeter gives primary stress to autonomous features of the capitalist process and says very little about the role of government in economic life. Where he does mention government, it is usually to scorn its perversity, as in his attack on Roosevelt's New Deal — though he regards government as pretty impotent. For him the driving force in economic life is entrepreneurship, which he regarded as having been taken over more or less completely by large firms. The emphasis on entrepreneurship is present in Schumpeter's earliest work on capitalist development written in 1911, and is obviously influenced by the ideas of Max Weber and Werner Sombart, which were popular at that time.

Schumpeter's big-wave chronology was rather similar to that of Kondratieff, though he gave each of the big waves a name and divided each wave into four phases rather than two.[24]

The main weaknesses of Schumpeter's long-wave theory (ignoring his failure to demonstrate their existence in the real world) are threefold: (1) he does not provide a persuasive explanation why innovation (and entrepreneurial drive) should come in regular waves rather than in a continuous but irregular stream, which seems a more plausible hypothesis for analysis concerned with the economy as a whole; (2) he makes no distinction between the lead country and the others, but argues as if they were all operating on a par as far as productivity level and technological opportunity is

TABLE 4.5. *Schumpeter's Long-Wave Chronology*

Prosperity	Recession	Depression	Revival
1. Industrial Revolution Kondratieff (cotton textiles, iron, and steam power)			
1787–1800	1801–13	1814–27	1828–42
2. Bourgeois Kondratieff (railroadization)			
1843–57	1858–69	1870–85	1886–97
3. Neomercantilist Kondratieff (electricity, automobiles, chemicals)			
1898–1911	1912–25	1925–39	?

concerned. Thus his waves of innovation are expected to affect all countries simultaneously; (3) he greatly exaggerates the scarcity of entrepreneurial ability and its importance as a factor of production.

Schumpeter developed his ideas on capitalist dynamics in another book published during the Second World War (*Capitalism, Socialism and Democracy*), which is not concerned with long waves but presents a breakdown theory for the capitalist system. This is rather paradoxical coming from an analyst who had such great faith in the robust character of capitalism, but his breakdown theory is socio-political rather than economic. He argued that there were four major forces destroying capitalism. In the first place, entrepreneurship was likely to be stifled by bureaucratization of management and decision-making in large firms. The second menace was the disincentive of progressive taxation and the increasing power of trade unions, which had already (he argued) retarded US recovery in the 1930s and was likely to become more stifling. The third threat came from the growing power of socialist ideas, and the fourth from the unpopularity of capitalism with intellectuals, who were continually engaged in denunciatory activities and harassments such as anti-trust suits.

Schumpeter's approach to long waves and the breakdown of capitalism has great fascination. It contains bold hypotheses and unsettling paradoxes, which gain in impact through his emotional detachment. His view of capitalist development is fatalistic, and he writes as if he were charting destiny. He dislikes most of what is happening in the real world, but does not advocate policies to remedy the predicted catastrophe. In fact, one is never sure with Schumpeter whether he is putting forward a specific hypothesis because he seriously believes it or because it is a stimulating illustration of his fundamentally dynamic and original conception of capitalist development.

Long-Wave Revivalists

The significant change in the momentum of economic growth after 1973 revived the notion of long rhythms in economic life and a number of new long-wave pundits emerged. Some of these are vulgarizers of past long-wave theories, which they invoke uncritically in support of a fashionable gloom about the future;[25] others deserve critical inspection even though I have not found much in their work to shake my scepticism about long waves as a systematic phenomenon affecting output. The two revivalists examined here are W. W. Rostow and E. Mandel.[26]

Rostow is concerned with 'Kondratieff' movements, essentially in the sense of swings in the terms of trade of primary producers against those selling industrial goods. Thus he refers to the 1951–73 period as the 'downswing' of a fourth Kondratieff, and the OPEC-inspired price increases as the upswing of a fifth Kondratieff, which poses particular problems because of demographic pressures in the developing world. Although I myself feel uncomfortable about calling the 1951–73 period a 'downswing', the Rostow thesis in itself appears fairly reasonable, and Arthur Lewis has shown the interest in explaining this facet of Kondratieff's work. Furthermore, Rostow produces 800 pages of empirical material to back his argument, in welcome contrast to some of his earlier work. However, he complicates his argument by embedding it in a loosely integrated framework that features neo-Schumpeterian surges of innovation in leading sectors, demand changes as economies work themselves through a hierarchy of stages, and a reiteration of his earlier erroneous belief that there was a short, sharp take-off in Western countries which was staggered in time. Like Schumpeter, Rostow has little time for broad aggregates such as GDP, which to my mind are the central indicators to be used in measuring acceleration or deceleration of growth.

Mandel approaches long waves from a rather different ideological position from Rostow — being an erudite Belgian Marxist of Trotskyite persuasion. He asserts that there are long swings, roughly fifty years in length, caused by surges of new technology. In each swing there are two phases. In the first phase profits rise as new technology is developed, and in the second phase profit rates fall as technical possibilities are exhausted. The timing, like the causality, is similar to Schumpeter's. The first (Industrial Revolution) wave was from the 1780s to 1847; the second, from 1847 to the 1890s, is attributable to a technological revolution dominated by 'machine production of steam motors'; the third, from the 1890s to 1939, is associated with the 'machine production of electric and combustion motors'; and the fourth, from 1940 to a future unspecified date, is associated with machine production of electronic motors and atomic energy. He suggests that the first phase of the fourth wave ended in 1967 and that we are now in the second phase. Unlike other writers in this vein, he does not refer to the waves as 'Kondratieffs', as he considers Kondratieff unoriginal as compared with van Gelderen (for whose work he has an exaggerated respect).

Mandel is mainly interested in theory and the empirical underpinning is very weak. He claims (p. 137) that 'economic historians are practically unanimous' in distinguishing expansions and recessions in the periods he uses, but the only justification he gives for this is an article by Hans Rosenberg published in 1943, which itself contained no empirical material and was written before quantitative economic history began. He also presents estimates of industrial production for the UK, Germany, and the USA and estimates of world trade to buttress his argument. These are not deviations from detrended moving averages, but compound rates of growth between the years specified (which vary by type of indicator).

Table 4.6 shows Mandel's indicators for the second-wave downswing, which he calls a period of 'pronounced depression', and for the third-wave upswing, which he calls a period of 'tempestuous increase in economic activity'. The figures do not bear out such a dramatic conclusion, particularly if one uses the alternative measures in the bottom half of the table, which are drawn from more recent sources but cover exactly the same periods and refer to the same concepts as those of Mandel.

Mandel considers that there have been stages as well as waves

TABLE 4.6. *Mandel's Evidence Scrutinized* (annual average compound growth rates)

		Second-wave downswing 'Pronounced depression'	Third-wave upswing 'Tempestuous increase in economic activity'
Mandel's indicators	UK industrial output	1.2 (1876–93)	2.2 (1894–1913)
	Germany industrial output	2.5 (1875–92)	4.3 (1893–1913)
	USA industrial output	4.9 (1874–93)	5.9 (1894–1913)
	World trade	2.2 (1870–90)	3.7 (1891–1913)
	Average	2.7	4.0
Mandel replicated	UK industrial output	1.4	2.4
	Germany industrial output	4.0	4.2
	USA industrial output	4.9	5.0
	World trade	3.4	3.5
	Average	3.4	3.8

Sources: First five rows from E. Mandel, *Late Capitalism*, New Left Books, London, 1975, pp. 141–2 (I have omitted his citation of Dupriez's 1947 estimates of world per capita output as these are much too shaky for serious use in this context). My indicators of industrial production including construction for the UK, Germany, and the USA are from W. A. Lewis, *Growth and Fluctuations 1870–1913*, Allen & Unwin, London, 1978, and for world trade volume from A. Maddison, 'Growth and Fluctuation in the World Economy 1870–1960', *Banca Nazionale del Lavoro Quarterly Review*, June 1962.

of development within the capitalist period; but, interestingly enough, although he calls his book 'late capitalism', he claims that this is not a new stage but merely a development within the second stage of imperialist monopoly-capitalism, which Lenin distinguished from the first phase of 'free competition'.

At first sight this restraint is puzzling, for Mandel frequently refers to features of 'late capitalism' that seem rather different from those that Lenin distinguished, e.g. the enhanced role of the state in the economy, the formal ending of colonialism, the importance of military spending, and the changed international power locus. The reason for Mandel's position is explained towards the end of his book, where he makes it clear that he wants to avoid being classified with the type of 'revisionist' who claims that there is a new era of state capitalism with a mixed economy that can 'suspend the internal economic contradictions of capitalism'.[27]

Thus there is no real connection between Mandel's stages of growth and his long waves. The latter are the fruit of more or less exogenous technological development, and do not have the policy-institutional flavour that Schumpeter conferred on his by calling one 'bourgeois' and another 'neo-mercantilist'.

Although I disagree with Mandel's conclusion that he has found empirical evidence for long waves, his theoretical position has interesting elements of originality, and his discussion of the intellectual history of this field is also more stimulating than many other accounts.

Conclusions on Long-Wave Theories

My basic conclusion is that the existence of regular long-term rhythmic movements in economic activity is not proven, although many fascinating hypotheses have been developed in looking for them. Nevertheless, it is clear that major changes in growth momentum have occurred since 1820, and some explanation is needed. In my view it can be sought not in systematic long waves, but in specific disturbances of an *ad hoc* character. Major system shocks change the momentum of capitalist development at certain points. Sometimes they are more or less accidental in origin; sometimes they occur because some inherently unstable situation can no longer be lived with but has finally broken down (e.g. the Bretton Woods fixed exchange rate system). I also feel that the

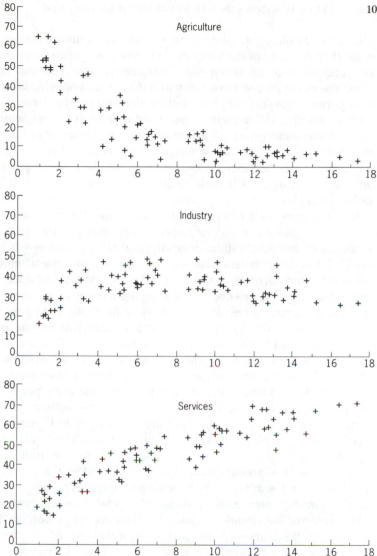

Graph 4.3 *Sectoral Distribution of Employment at Different Levels of Real Per Capita GDP, 1870–1987*

Vertical scale shows proportions of employment in agriculture, forestry, and fishing (top graph), industry including mining, manufacturing, utilities, and construction (middle graph), and services (bottom graph). The horizontal scale shows levels of GDP per capita ('000 US$ in 1985 prices), using data for 1870, 1950, 1960, 1973, and 1987 for 16 countries.

institutional-policy mix plays a bigger role in capitalist development than do many of the long-wave theorists. A system shock will produce the need for new policy instruments, and these are not always selected on the most rational basis; or they may require a long period of experiment before they work properly. There may also be conflicts of interest within and between countries which prevent the emergence of efficient policies. Hence there may well be prolonged periods in which supply potential is not fully exploited. Some of these problems are faced in Schumpeter's analysis but usually as if their solution were a matter of destiny rather than choice.

It is also important to keep in mind that capitalist development since 1820, though it has a certain unity because the growth momentum has lain within distinctly higher limits than earlier epochs, has nevertheless seen big changes in the character of economic life which were bound to influence the type of fluctuations that were experienced. These changes have to be kept in mind in constructing any general theory of fluctuations or phases. One of these is the change in the structure of production and employment that has resulted from increased levels of income and changed patterns of demand and productivity. In 1820, agriculture characteristically employed well over half of the labour force in these countries, whereas the average has now fallen to 6 per cent. Agriculture was and still is subject to erratic fluctuations in output owing to weather, and its products are generally sold in flexprice markets in which prices go down as well as up. This erratic element in economic life is now much smaller than it used to be. Industrial employment was probably about a quarter of the total around 1820 and rose towards a peak of somewhere round 50 per cent in most countries. Hence the process of capitalist development is often referred to as industrialization, with its first phase as the 'Industrial Revolution', and particular weight is often placed on industrial production as an index of growth. However, the industrial share in employment has been on the decline for the past thirty years, and has now regressed closer to the 1820 proportion than to its peak level. The big long-run gains have been in services, which had perhaps a fifth of total employment in 1820 against two-thirds now. It was in the industrial sector that the business cycle was most marked in terms of stock-output supply adjustments and fluctuations in demand, but in the service sector both demand and supply are more stable, and this has dampened the amplitude of fluctuations in GDP.

A second major change in economic life has been the growing role of government. In 1820 government consumption was typically less than 10 per cent of GDP, but the proportion has now doubled. In addition, government intervenes on a massive scale to operate a vast network of social transfers, which change the distribution of income and the pattern of private spending. Total government spending is now nearly half of GDP. Finally, the government regulatory role in the economy has greatly increased. One result of the latter is that the stability of financial institutions has improved. Before the Second World War, depressions were often reinforced by major bank failures, but these are now rarer and their impact is cushioned, though the potential for such disturbances still exists. As a result of these changes, government exercises both a propulsive and a compensatory role in economic life, which generally operates to stabilize the expenditure and income flow, and the aspirations of governments to act as managers of economic destiny have greatly increased.

There are also other changes to keep in mind when developing hypotheses intended to cover the whole capitalist period. An important one is the change in the average size of firms, and the growth of trade unions to represent the interests of workers. Hence, the atomized market paradigm is no longer very relevant in wage and price fixing, which explains some of the changes that have occurred in price behaviour. Another is the character of the international linkages between countries, which have varied a good deal over time and which have been the most exposed to system shock. One fundamental aspect of this is the nature of the international monetary system, which has a major impact on the type of policy weapons used domestically. Others are the level of trade, migration, and capital movements, and the scope for international transfers of technology.

PHASES OF GROWTH

Although I find no convincing evidence in the work of Kondratieff, Kuznets, and Schumpeter to support the notion of regular or systematic long waves in economic life, there have nevertheless been significant changes in the momentum of capitalist development. In the 170 years since 1820 one can identify separate phases which have meaningful internal coherence in spite of wide variations in individual country performance within each of them. Phases are identified, in the first instance, by inductive analysis and iterative inspection of empirically measured characteristics. In order to illustrate trends, cycles, and phases, estimates are presented for as many individual years as possible, including war years. I have also aggregated movements for the sixteen countries as a whole, showing both weighted and unweighted averages. For many purposes the unweighted average is the most relevant indicator of the characteristic experience of these countries, because countries are our basic unit of analysis. For some purposes a weighted average is a useful supplement, but it should not be forgotten that the USA has a very large weight in such averages, particularly for the twentieth century. For many indicators, information is poor before 1870. Hence our systematic presentation of data is restricted to the period following 1870, but the available evidence suggests that in most respects the 1820–70 experience was similar to that in 1870–1913.

Table 4.7 gives a summary view of the amplitude of annual changes in GDP, which is our preferred measure of aggregate output for the sixteen countries taken together. Table 4.8 gives a synoptic view of the incidence of recession by year, and by country. The biggest interruptions to growth occurred in the 1930–2 depression, and in the 1945–6 period of demobilization, dismemberment, defeat, and victory. All other disturbances had a much milder impact on output, including those of the First World War and its aftermath. The aggregate stability in the collective output of the group in peacetime has been quite impressive. In the forty-three years from 1870 to 1913, there were only three years of recession in aggregate output, in the twenty-five years 1947–73 none, and in 1974–89 two. However, it is clear from Table 4.8 that individual countries have been much more unstable than the group as a whole (particularly before 1913). The cyclical experience of individual countries has not normally been synchronized, but rather compensatory. Cyclical experience has been synchronized only when they have been subjected to 'system shocks' such as wars, or the collapse of long-standing international payments mechanisms.

TABLE 4.7. *Year-to-Year Percentage Change in Aggregate GDP of the Sixteen Countries, 1871–1989*

Year	Change	Year	Change	Year	Change
1871	2.8	1914	−5.9	1950	7.9
1872	4.1	1915	3.1	1951	8.5
1873	1.9	1916	8.7	1952	4.1
1874	3.5	1917	−2.4	1953	4.7
1875	3.0	1918	1.7	1954	1.7
1876	−0.7	1919	−1.4	1955	6.1
1877	1.8	1920	0.0	1956	3.5
1878	1.9	1921	−1.0	1957	3.1
1879	2.7	1922	6.7	1958	1.1
1880	6.8	1923	6.3	1959	5.7
1881	2.8	1924	4.9	1960	4.8
1882	4.4	1925	3.8	1961	4.4
1883	1.7	1926	3.6	1962	5.4
1884	1.0	1927	2.5	1963	4.8
1885	0.3	1928	3.2	1964	6.5
1886	2.4	1929	4.3	1965	5.1
1887	3.8	1930	−5.7	1966	5.0
1888	1.2	1931	−6.4	1967	3.6
1889	3.8	1932	−6.6	1968	5.6
1890	2.4	1933	1.5	1969	5.2
1891	1.8	1934	5.1	1970	3.1
1892	3.4	1935	5.6	1971	3.4
1893	−1.1	1936	8.6	1972	5.0
1894	1.9	1937	5.4	1973	5.7
1895	5.0	1938	−0.1	1974	0.5
1896	1.0	1939	7.1	1975	−0.4
1897	3.7	1940	3.2	1976	4.7
1898	4.9	1941	8.5	1977	3.8
1899	5.0	1942	9.1	1978	4.3
1900	2.4	1943	9.4	1979	3.3
1901	4.2	1944	2.4	1980	1.3
1902	0.8	1945	−8.1	1981	1.8
1903	3.6	1946	−11.1	1982	−0.4
1904	0.5	1947	1.6	1983	2.9
1905	4.5	1948	5.5	1984	5.0
1906	7.0	1949	3.6	1985	3.5
1907	3.1			1986	2.7
1908	−3.6			1987	3.4
1909	6.6			1988	4.4
1910	1.2			1989	3.5
1911	3.9				
1912	4.1				
1913	3.3				

TABLE 4.7. (Cont.)

Source: Weighted estimates derived from Appendix A. Figures exclude Japan before 1885, the Netherlands before 1900, and Switzerland before 1899. For the First World War there were some gaps in data for Australia, Belgium, and Switzerland for which rough estimates were made; this is also true for Belgium, 1939–47, Switzerland, 1944–6, and Japan, 1945–6; see Appendix A for the interpolations. The figures are adjusted to exclude the impact of territorial change.

My method of distinguishing phases of development is quite simple. It involves collecting annual time series for major indicators of economic activity for the sixteen countries in as complete and comparable a form as possible, and by inspection of these and graphs derived from them, identifying fundamental turning-points in growth momentum, and trying to establish growth and cyclical behaviour patterns that differ significantly between phases. The technique is not unlike that of the NBER in its attempt to identify reference cycles, and in particular does not involve elaborate decomposition of time series into different kinds of oscillatory movement. Simple techniques such as this are almost inevitable in handling information for sixteen countries, where each series, if it were available for the full 170 years, would involve more than 2,700 readings. Furthermore, it is necessary in this kind of comparative historical work to be very careful in making adjustment to enhance the comparability of the basic data. There is some danger in overprocessing results drawn too mechanically from such data.

In analysing the sequence of phases, the first problem is one of periodicity. Table 4.7 suggests that the period 1870–1913 has a certain unity in that growth was moderate and interrupted by recession, but not subject to the extreme shocks which struck three times between 1914 and the 1940s. There was also something special about the unprecedented secular boom which started in 1947 and ended in 1973. Evidence of various kinds suggests that the nature of the growth process changed after 1973. I have therefore distinguished four phases: 1870–1913, 1913–50, 1950–73, and 1973 onwards. However, my hunch, based on partial indicators for a few of the countries, is that the first phase can be extended to include 1820–1913 as a whole.

Kuznets postulates five minimum requirements for acceptable stages of growth:[28] (1) they must be identified by characteristics

TABLE 4.8. *Incidence of Recessions, 1870–1989* (years and countries in which GDP fell)

Year	No. of falls	Countries affected	Year	No. of falls	Countries affected	Year	No. of falls	Countries affected
1871	2	AG	1914	12	ATBCLFGIJNSE	1950	0	
1872	2	CI	1915	8	ATBDLFGS	1951	1	D
1873	3	TDF	1916	2	TS	1952	2	BK
1874	2	IE	1917	9	TDLFNWSZE	1953	0	
1875	3	BCS	1918	10	TBCDLFNWSZ	1954	2	CE
1876	5	ACFGI	1919	6	ATCGIK	1955	1	D
1877	4	DLGS	1920	5	CIJKE	1956	0	
1878	5	CLFWS	1921	9	CDFIWSZKE	1957	0	
1879	5	TFGSK	1922	1	J	1958	5	BNWZE
1880	1	G	1923	2	TG	1959	0	
1881	1	I	1924	1	W	1960	0	
1882	4	ACLW	1925	1	D	1961	0	
1883	3	FIW	1926	2	AK	1962	0	
1884	1	F	1927	3	AFI	1963	0	
1885	5	TCFSK	1928	1	A	1964	0	
1886	1	S	1929	5	ABCLG	1965	0	
1887	0		1930	13	ATBCLFGIJNZKE	1966	0	
1888	5	TFIJE	1931	14	ATBCLFGINWSZKE	1967	1	G
1889	4	TCFI	1932	10	TBCDFGNSZE	1968	0	

TABLE 4.8. (Cont.) *Incidence of Recessions, 1870–1989* (years and countries in which GDP fell)

Year	No. of falls	countries affected	Year	No. of falls	Countries affected	Year	No. of falls	Countries affected
1890	1	A	1933	5	TCINE	1969	0	
1891	4	LGIJ	1934	3	BFN	1970	1	E
1892	5	ACLIK	1935	2	FZ	1971	0	
1893	3	ACE	1936	0		1972	1	S
1894	2	IE	1937	0		1973	0	
1895	3	ACF	1938	4	BFNE	1974	4	DJKE
1896	2	JE	1939	2	LZ	1975	10	TBDFGINZKE
1897	3	AFI	1940	8	TBDLFNWS	1976	1	Z
1898	0		1941	6	BDFINZ	1977	1	S
1899	2	LJ	1942	8	TBFIJNWZ	1978	0	
1900	1	K	1943	6	BFINWZ	1979	0	
1901	6	ALFGNS	1944	8	ALFIJNWK			

Year		Countries
1902	5	LFIJS
1903	2	WK
1904	3	NWE
1905	1	J
1906	0	
1907	0	
1908	5	CNSKE
1909	3	TJS
1910	2	FI
1911	0	
1912	0	
1913	0	

Year		Countries
1945	10	ATCDLGIJKE
1946	5	ACGKE
1947	2	KE
1948	0	
1949	1	Z

Year		Countries
1980	3	DKE
1981	6	TBDNSK
1982	4	CGZE
1983	0	
1984	0	
1985	0	
1986	0	
1987	1	D
1988	1	D
1989	0	

Country code:

A	Australia	D	Denmark	I	Italy	S	Sweden
T	Austria	L	Finland	J	Japan	Z	Switzerland
B	Belgium	F	France	N	Netherlands	K	UK
C	Canada	G	Germany	W	Norway	E	USA

Source: Appendix A. Figures exclude Japan before 1885, the Netherlands before 1990, and Switzerland before 1899.

that can be verified or quantified; (2) the magnitude of these characteristics must vary in some recognizable pattern from one phase to another ('stages are presumably something more than successive ordinates in the steadily climbing curve of growth. They are segments of that curve with properties so distinct that separate study of each segment seems warranted'); (3) there should be some indication of when stages terminate and begin and why; (4) it is necessary to identify the universe to which the stage classification applies; (5) finally, Kuznets requires that there be an analytic relation between successive stages, which, optimally, would enable us to predict how long each stage has to run. This seems to me too deterministic. It suggests that movements between successive stages are more or less ineluctable. As I cannot meet Kuznets's fifth requirement, my periods are 'phases' rather than 'stages'.

My growth phases fulfil the first four of Kuznets's requirements as explained below.

1. The phases are identified by eight simple indicators showing both growth and cyclical characteristics: rate of growth of out-

TABLE 4.9. *Growth Characteristics of Different Phases, 1870–1989* (arithmetic average of figures for the individual countries)

Phases	GDP	Annual average compound growth rates		
		GDP per head of population	Tangible reproducible gross non-residential fixed capital stock[a]	Volume of exports
I 1870–1913	2.5	1.4	3.4	3.9
II 1913–50	2.0	1.2	2.0	1.0
III 1950–73	4.9	3.8	5.8	8.6
IV 1973–89	2.6	2.1	4.2	4.7

[a] refers to six countries, first period is 1890–1913, last one is 1973–87.
Sources: Tables 3.1, 3.2, 5.4, 3.15.

TABLE 4.10. Cyclical Characteristics of Different Phases, 1870–1989 (arithmetic average of figures for the individual countries)

Phase	Maximum peak–trough fall in GDP (or smallest rise)	Maximum peak–trough fall in export volume	Average unemployment rate (% of labour force)	Average annual rise in consumer prices
I 1870–1913	−5.6	−18.2	4.5[a]	0.4
II 1920–38	−12.4	−36.5	7.5	−0.7[b]
III 1950–73	+0.2	−7.0	2.6	4.1
IV 1973–89	−1.8	−7.6	5.7	7.3

[a] UK and USA, 1900–13.
[b] 1924–38 for Austria and Germany, 1921–38 for Belgium.

Sources: Tables 4.1, 4.2, C.6, and 6.3.

TABLE 4.11. *System Characteristics of Different Phases*

Governmental policy stance on unemployment/price stability trade-off	Nature of international payments system	Labour market behaviour	Degree of freedom for international trade	Degree of freedom for international factor movements
I: *1870–1913 'Liberal Phase'*				
No concern with unemployment	Gold (sterling) standard with rigid exchange rates	Weak unions; wages had some downward flexibility	Very free. No QRs or exchange restrictions. Tariffs the only barrier	More or less complete freedom
II: *1913–50 'Beggar-Your-Neighbour' Phase*				
Concern with price and exchange stability leads to conscious acceptance of large-scale unemployment	Gold standard restored at nostalgic parities, quarrels over government debt 1931 system collapse followed by moveable peg	Governments enforce downward wage flexibility	QRs and exchange restrictions widespread. Tariffs raised substantially	Severe controls on both capital and labour

III: *1950–73 'Golden Age'*				
Priority given to full employment	Fixed (but not rigid) exchange rates with large international credit facilities	Strong unions, no downward wage flexibility	Very strong move towards freer trade and customs unions	Gradual and substantial freeing of both labour and capital movements
IV: *1973 onwards 'Phase of Cautious Objectives'*				
Priority given to price stability	System collapse followed by floating rates and growing area of stability in EMS	Weakened unions	Free trade maintained	Freedom for capital movements augmented, labour movement restricted

put, output per head, capital stock and export volume, cyclical variations in output and exports, levels of unemployment, and rate of price increase. These are the conventional macroeconomic indicators one might use for growth accounting or conjunctural monitoring. The results are shown in very aggregative form in Tables 4.9 and 4.10. Each phase also has five non-quantifiable 'system characteristics', by which I mean the basic policy approaches and institutional environment that condition growth performance. These include the government approach to demand management (i.e. the kind of trade-off that is made between unemployment and inflation), the bargaining power of labour, the degree of freedom for trade and international factor movements, and the character of the international payments mechanism. Changes in these between periods are summarized in Table 4.11.

2. Most of the characteristics are systematically different in the four phases identified. Generally, they are most favourable in phase III, second-best in phase IV, third-best in phase I, and worst in phase II. The exceptions to the second-best rating are the pace of price increase, where phase IV is worst; and unemployment, where it is second-worst.

3. There is room for argument as to which years are terminal for demarcation purposes, particularly as the use of annual data means that the periodicity has to be rather precise. I explained in Chapter 3 why I picked 1820 as the starting-point for capitalist development; 1913 is clearly the last year of phase I, which ended with the outbreak of the First World War; and 1950 was chosen as a point where recovery from the Second World War was more or less completed in terms of recovery of the previous peak in output for the sixteen countries as a whole. However, five countries did not pass their wartime output peaks until 1953 (Austria, Germany, Japan, UK, and USA) so one might well argue that 1953 rather than 1950 should mark the beginning of the post-war golden age. On the other hand, there is a case for starting in 1948, which is when the ground rules for international co-operation within the capitalist group were set up by the Marshall Plan; so 1950 seems a reasonable compromise. It should be noted that use of 1948–73 or 1953–73 instead of 1950–73 would not affect the analysis seriously — the third phase would still be a period of secular boom on an unparalleled scale, and the second would still have the worst performance.

4. The emergence of a fourth phase after 1973 is quite clear. The 1974–5 and 1980–2 recessions affected virtually all sixteen countries. They were by far the biggest breaks in the post-war growth momentum. The grounds for treating the post-1973 period as a new phase include price, unemployment, and output behaviour, changes in the international monetary system, in government policy concerning the level of demand, in expectations in the labour market, and greater openness of capital markets. The economic system behaves in a different way, which has created major new tasks for economic policy, and makes it more difficult to reconcile different policy objectives.

Recognition of the phase phenomenon forces consideration of factors operating for these countries as a whole. The interrelatedness of their economies limits the options which each is able or willing to pursue. Hence each phase has demonstrated a distinctive orbit, which puts some constraints on feasible national trajectories of growth and change. These constraints must be part of the explanation for the surprising generality of the phase phenomenon.

The main conclusions I would draw about major fluctuations in the momentum of capitalist development are as follows.

1. There are distinct phases of economic performance, each with its own momentum.
2. Phases of growth are not ineluctable, and within each there is considerable scope for variation in country performance; but the policy-institutional framework and policy attitudes characteristic of each phase have had a striking distinctiveness and generality of acceptance. The expectations of economic agents about growth and inflation have also had distinctive characteristics which differ between phases.
3. The move from one phase to another has been caused by system shocks. These may well be due to a predictable breakdown of some basic characteristic of a previous phase, but the timing of the change is usually governed by exogenous or accidental events which are not predictable.
4. A more specific conclusion is that developments since 1973 represent a new phase and not just a temporary interruption of phase III.

5. The present phase generally ranks as second-best. Performance is well below that in phase III in almost all important respects, but the economy has been a good deal more stable in real terms than before 1950, and the growth of output per capita is significantly better than in the first two phases.

NOTES ON CHAPTER 4

1. See M. Bronfenbrenner (ed.), *Is the Business Cycle Obsolete?*, John Wiley, New York, 1969, and the comments of R. M. Solow, *Economic History Review*, Dec. 1970: 'The old notion of a fairly regular self-sustaining "business cycle" is not very interesting anymore. Today's graduate students have never heard of Schumpeter's apparatus of Kondratieffs, Juglars, and Kitchins, and they would find it quaint if they had.'
2. See C. Juglar, *Des crises commerciales et de leur retour périodique en France, en Angleterre et aux États Unis*, Kelley (reprint), New York, 1967, p. 256.
3. See M. von Tugan-Baranowsky, *Studien zur Theorie und Geschichte der Handelskrisen in England*, Fischer, Jena, 1901, which develops underconsumptionist explanations of the business cycle.
4. See W. L. Thorp, *Business Annals*, NBER, New York, 1926; A. F. Burns and W. C. Mitchell, *Measuring Business Cycles*, NBER, New York, 1947, pp. 78–9. See also W. C. Mitchell, *Business Cycles: The Problem and Its Setting*, NBER, New York, 1930, for an excellent history of cyclical analysis.
5. See Burns and Mitchell, *Measuring Business Cycles*, p. 270, who state the reasons for not eliminating trend, with which I entirely agree: 'cyclical fluctuations are so closely interwoven with these secular changes in economic life that important clues to the understanding of the former may be lost by mechanically eliminating the latter. It is primarily for this reason that we take as our basic unit of analysis a business cycle that includes that portion of secular trend falling within its boundaries.'
6. In the period 1889–1978, the NBER recorded twenty-one reference cycles, the industrial production index showed fifteen recessions, and GDP thirteen. The average amplitude of GDP recessions was a 6.5 per cent fall, and of industrial production, 13.3 per cent. Before 1889 the GDP index for the USA contains too heavy an element of interpolation to be used for cyclical analysis.
7. Estimates are available for Denmark, France, and Sweden for 1820–70, and the UK for 1830–70. During these periods these countries showed maximum peak–trough GDP falls of 5.6, 11.5, 9.7, and 7.0 per cent respectively, i.e. an average of 8.5 per cent.
8. See N. D. Kondratieff, 'Die langen Wellen der Konjunktur', *Archiv für Sozialwissenschaft and Sozialpolitik*, Dec. 1926, pp. 573–609.
9. The most sophisticated discussion of the Kondratieff wave in prices for the 1870–1913 period is contained in W. A. Lewis, *Growth and Fluctuations 1870–1913*, Allen & Unwin, London, 1978, which examines whether prices influenced output movements or output influenced prices. Lewis also discusses the role of gold production. His conclusion is that the global price movement in this period was most strongly influenced by US agricultural production. Although Lewis uses personalized nomenclature for various cycles and waves, as Schumpeter also did, he does not in fact endorse the idea of Kondratieff waves as a non-monetary phenomenon on an international scale.
10. It is rather odd that Kondratieff eliminated the population component in which the Kuznetsians have found the best evidence for their own long-wave analysis.
11. See Kondratieff, 'Die langen Wellen', pp. 586 (graphs) and 607–9 for the data and trend formulae. Kondratieff's graph for coal should be compared with the minor ripples shown in that of S. S. Kuznets, *Secular Movements in Produc-*

tion and Prices, Houghton Mifflin, Boston, 1930, p. 124, which shows proportionate deviations from a trend calculated from a different formula.

12. See G. Garvy, 'Kondratieff's Theory of Long Cycles', *Review of Economic Statistics*, Nov. 1943, for an excellent review of Kondratieff's work and account of his Soviet critics.

13. It is sometimes suggested that Kondratieff's approach was no advance on ideas put forward by van Gelderen under the pseudonym J. Fedder, 'Springvloed', *De Nieuwe Tijd*, Fortuyn, Amsterdam, 1913. In fact, he may not have proved much more than van Gelderen — i.e. that there are long swings in the general price level — but in terms of analytic framework and statistical technique, what Kondratieff offered was distinctly novel.

14. See S. S. Kuznets, *Secular Movements in Production and Prices*, Kelley (reprint), New York, 1967.

15. Kuznets presented twenty-three indicators for the USA, of which seventeen were commodities with both price and quantity data, five were financial indicators, and the last was the general price index. For the UK he had nine indicators, France and Germany eight each, Belgium five, Canada and Japan two each, Australia and Argentina one each.

16. See Burns and Mitchell, *Measuring Business Cycles*, p. 428: 'Kuznets did not draw up a list of dates showing the peaks and troughs of his "secondary secular variations". In attempting to determine such a chronology from his American series, we found their turning points so widely dispersed that we could have little confidence in any list we ourselves might extract.'

17. See S. Kuznets, *Economic Development and Cultural Change*, Oct. 1956, p. 50. This article was rewritten and published as ch. 1 of *Economic Growth of Nations*, Harvard, 1971, where Kuznets dropped his aggregate chronology.

18. 'Long Swings in Population Growth and Related Economic Variables', reprinted in S. Kuznets, *Economic Growth and Structure*, Heinemann, London, 1965, pp. 328–78. See also S. Kuznets, *Capital in the American Economy*, Princeton, 1961, ch. 2, 7, 8, and 9.

19. The others include B. Thomas, *Migration and Economic Growth*, Cambridge, 1954; J. G. Williamson, *American Growth and the Balance of Payments: A Study of the Long Swing*, Chapel Hill, 1964; R. A. Easterlin, *Population, Labor Force, and Long Swings in Economic Growth*, NBER, New York, 1968.

20. See *Historical and Comparative Rates of Production, Productivity and Prices*, Part 2 of *Hearings on Employment, Growth, and Price Levels*, Joint Economic Committee, US Congress, Apr. 1959, pp. 411–66; and M. Abramovitz, *Thinking About Growth*, Cambridge, 1989, which is a collection of his major articles on this and related topics.

21. See J. Kitchin, 'Cycles and Trends in Economic Factors', *Review of Economic Statistics*, Jan. 1923, pp. 10–16.

22. Page references are to J. A. Schumpeter, *Business Cycles*, McGraw-Hill, New York, 1939.

23. See J. A. Schumpeter, *Capitalism, Socialism and Democracy*, Allen & Unwin, London, 1943, p. 64.

24. In fact, Schumpeter was not very explicit on his chronology, which we owe to Kuznets's exegesis after consultation with Schumpeter; see S. Kuznets, 'Schumpeter's Business Cycles', *American Economic Review*, June 1940, pp. 250–71, for a highly sceptical assessment.

25. See J. J. van Duin, *De lange golf in de economie*, Van Gorcum, Assen, 1979, who is an eclectic revivalist, rather cavalier with the few empirical facts he presents; J. W. Forrester, 'Growth Cycles', *De Economist*, 1977, pp. 525–43, produces long waves with no data. J. S. Goldstein, *Long Cycles*, Yale, 1988,

presents a dizzying survey of most of the work in Western languages on long waves. S. M. Menshikov and L. A. Klimenko, *Dlinnie Volni v Ekonomike*, Mezdunarodnie Otnoschenia, Moscow, 1989, present a Soviet view. Well-documented scepticism about long waves can be found in W. H. Schröder and R. Spree (eds.), *Historische Konjunkturforschung*, Klett Cotta, Stuttgart, 1981, S. Solomou, *Phases of Economic Growth, 1850–1973*, Cambridge, 1987, and J. P. G. Reijnders, *The Enigma of Long Waves*, Ph.D. thesis, Groningen, 1988.

26. See W. W. Rostow, 'Kondratieff, Schumpeter and Kuznets: Trend Periods Revisited', *Journal of Economic History*, Dec. 1975, which contains the essentials of the approach in his *The World Economy*, Macmillan, London, 1978. See E. Mandel, *Late Capitalism*, New Left Books, London, 1975. In addition to these two authors, there are elements of originality in G. Mensch, *Das technologische Patt*, Frankfurt, 1975, who has a Schumpeterian-type approach with a detailed catalogue of different types of innovation. He considers that the clustering of innovations determines the tempo of capitalist performance, and that the 1970s slow-down is due to a shortage of exploitable innovations. Mensch has some interesting ideas about lags in application of inventions, but lapses frequently into apocalyptic sermonizing. He presents almost no evidence on the variations in the pace of macroeconomic performance which he is presumably trying to explain, and nowhere makes the leader–follower dichotomy, which is fundamental in analysis of the diffusion of innovation.

27. Mandel cites several examples of this type of Marxist revisionism, of which the best example in my view is John Strachey, *Contemporary Capitalism*, Gollancz, London, 1956.

28. See S. Kuznets in W. W. Rostow (ed.), *The Economics of Take-Off into Sustained Growth*, Macmillan, London, 1963.

ACCELERATION, SLOW-DOWN, AND CONVERGENCE AFTER 1950

This chapter tries to answer four questions about growth performance since 1950: (1) Why was there such a general acceleration up to 1973? (2) Why did growth slow down so generally and so emphatically thereafter? (3) Why did all fifteen follower countries converge so markedly towards US levels of performance over the whole period? (4) Why did the performance of the lead country deteriorate so sharply after 1973?

The reality of acceleration and slow-down emerges dramatically from the four indicators in Table 5.1, which show differences in growth rates for successive periods. For GDP, GDP per capita, and labour productivity, all the entries on the left-hand side of the table are positive, i.e. there was acceleration in 48 out of 48 possible cases. The acceleration was biggest in Austria, Germany, Italy, and Japan and smallest in the USA, Canada, Switzerland, Norway, Sweden, and Australia. This country distribution suggests the influence of different degrees of recovery from the war, but the extent of the acceleration over all previous experience, and its prolonged character, means that there are other more fundamental reasons for the dynamism. In terms of population growth, the acceleration was much more modest. In this dimension, the element of recovery — from the war experience of deaths and fertility losses — was much stronger.

After 1973, the phenomenon of substantial and general deceleration is also clearly evident. It is manifest in 47 of the 48 output and productivity indicators. The slow-down was most marked in Japan but was very noticeable in Austria, France, Germany, Italy, the Netherlands, and Switzerland. The demographic slow-down after 1973 was also substantial in all the countries; this was probably not closely connected with the slow-down in real income growth, but was a reassertion of the much longer term decline in birth rates shown in Graph 3.2, which the post-war baby boom interrupted.[1]

TABLE 5.1. *Growth Acceleration and Slow-down after 1950* (differences in annual average compound growth rate)

	Acceleration from 1913–50 to 1950–73				Slow-down from 1950–73 to 1973–89			
	Population	GDP	GDP per capita	Labour productivity	Population	GDP	GDP per capita	Labour productivity[a]
Australia	0.8	2.5	1.7	1.2	−0.8	−1.6	−0.7	−0.9
Austria	0.3	5.1	4.7	5.0	−0.4	−2.9	−2.5	−2.2
Belgium	0.2	3.1	2.8	3.0	−0.4	−2.0	−1.5	−1.4
Canada	0.5	2.0	1.4	0.5	−1.0	−1.5	−0.4	−1.1
Denmark	−0.4	1.3	1.6	2.5	−0.6	−2.1	−1.5	−2.5
Finland	−0.1	2.2	2.4	2.9	−0.3	−1.8	−1.6	−3.0
France	0.9	3.9	2.9	3.1	−0.5	−2.7	−2.2	−1.8
Germany	1.7	4.6	4.2	4.9	−0.9	−3.8	−2.9	−3.3
Italy	0.0	4.1	4.2	3.8	−0.4	−2.7	−2.4	−3.2
Japan	−0.1	7.1	7.1	5.8	−0.4	−5.4	−4.9	−4.5
Netherlands	−0.1	2.3	2.3	3.5	−0.6	−2.7	−2.0	−2.4
Norway	0.0	1.2	1.1	1.7	−0.4	−0.1	0.4	−0.7
Sweden	0.0	1.3	1.2	1.6	−0.4	−2.0	−1.5	−2.8
Switzerland	0.9	1.9	1.0	0.6	−1.2	−3.2	−2.1	−2.1
UK	0.2	1.7	1.7	1.6	−0.4	−1.0	−0.7	−0.9
USA	0.2	1.2	0.6	0.1	−0.4	−0.9	−0.6	−1.5
Average	0.3	2.8	2.6	2.6	−0.6	−2.3	−1.7	−2.1

[a] Slow-down from 1950–73 to 1973–87.

Sources: Derived from Tables 3.1, 3.2, 3.3, and 3.7.

TABLE 5.2. *Rate of Divergence/Convergence of 15 Follower Countries towards Levels of Productivity in the Lead Country, 1890–1987* (annual average compound growth rates)

	1890–1913	1913–50	1950–73	1973–87
Australia	−0.33	−0.88	0.20	0.75
Austria	−0.49	−1.51	3.38	1.68
Belgium	−1.26	−1.00	1.86	2.13
Canada	0.69	−0.02	0.42	0.71
Denmark	−0.06	−0.79	1.62	0.53
Finland	0.01	−0.17	2.70	1.15
France	−0.48	−0.49	2.46	2.09
Germany	−0.36	−1.36	3.40	1.53
Italy	−0.20	−0.47	3.24	1.49
Japan	−0.49	−0.58	5.02	2.06
Netherlands	−1.07	−1.10	2.26	1.32
Norway	−0.52	0.04	1.74	2.38
Sweden	−0.31	0.33	1.89	0.59
Switzerland	−0.82	0.27	0.77	0.12
UK	−1.10	−0.83	0.71	1.28
Arithmetic average	−0.43	−0.57	2.11	1.32

Source: Appendix C, Table C.11, measuring rate of change in labour productivity level *vis-à-vis* the USA in the time periods specified. Divergence is shown by negative figures, convergence by positive magnitudes.

Table 5.2 shows the universality and scope of the post-war convergence of productivity levels in all fifteen follower countries in the post-war period. In 30 entries out of 30 since 1950 it is very clear. It was strongest in 1950–73, and generally slower thereafter. However, five countries (Australia, Belgium, Canada, Norway, and the UK) had faster convergency after 1973.

These four decades of post-war convergence are all the more striking because they contrast with a general pattern of divergence in the eight decades from 1870 to 1950. For these earlier periods we have 25 out of 30 cases of divergence. There were only 5 cases of convergence (Canada and Finland, 1890–1913;

Norway, Sweden, and Switzerland in 1913–50), and they were mild by post-1950 standards.

In analysing productivity developments since 1950 we must obviously try to establish why follower countries did so much better than the productivity leader, the USA. If the lead country had also experienced a significant post-war acceleration in growth, one might assume that there had been a quickening pace of progress at the technological frontier. As the USA did not have such an experience, none of the following growth accounts assume that embodied technical progress accelerated in the post-war period. The poor productivity performance and the low residual in the growth accounts of the USA since 1973 (see Tables 3.13 and 5.19) suggest rather that the pace of technological advance may have slowed down after 1973.

A good part of the productivity 'miracle' which occurred in the 1950–73 period was due to a favourable concatenation of political and policy circumstance (analysed in Chapter 6 below). High levels of demand maintained full employment and encouraged investment to rise rapidly. Capital stocks rose at an unprecedented pace. Labour moved from low productivity activities to more profitable employment. Barriers to international trade were removed, and international specialization improved.

The European countries and Japan were able to seize these opportunities because they had high levels of skill and education, and the institutional capacity to raise and allocate large capital resources. They were particularly successful in liberalizing international trade and capital flows, and in mitigating the business cycle. In these respects they were better placed for a catch-up effort than Third World countries, whose economic distance from the USA was greater, but whose human capital and institutional heritage were weaker, whose trade policy was more restrictive, and whose macropolicy was often more erratic.

The real mystery of the post-1973 slow-down is the sharp deceleration of productivity growth in the lead country, the USA. In the follower countries there is no mystery, and their slow-down had some elements of inevitability. As the productivity gap between the followers and the lead country narrowed, the scope for easy growth by replicating lead country technology was reduced. A greater burden of innovation fell on the followers, the profitability of high levels of investment faltered. Capital stock ceased to grow as fast. Similarly, the scope for easy structural change waned, because the relative size of low productivity sectors such as agriculture had fallen. Once the major barriers to trade were removed, further specialization gains accrued more slowly. All these predictable developments occurred, though the process of convergence on the leader still continues, the rate of capital accumulation is still high by historical standards, and the rate of productivity growth in the follower countries still exceeds the historical experience of 1820–1950.

However, there were other forces at work which help to explain why the productivity slow-down was so sharp and general after 1973 and why countries have not fulfilled their growth potential to the same degree as they did in 1950–73.

In the 1970s, the economic system of these countries suffered several major shocks: the collapse of the Bretton Woods fixed exchange system, the twelvefold increase in oil prices as a result of the two OPEC shocks, and the significant acceleration in the upward movement of other prices. These required major changes in the discretionary weapons of macroeconomic policy and the rules by which central banks and finance ministries operated. They also brought changes in the expectations of the private sector and in the behaviour of entrepreneurs and trade unionists. These 'shock' elements influenced post-1973 productivity in a temporary way. In addition, the general switch from Keynesian-type to more cautious macropolicies acted as a longer-term dampener. In the golden age high and stable levels of demand and low real interest rates produced buoyant attitudes. Instead of worrying about investment risks, entrepreneurs became more aware of the consequences of not investing: e.g. lack of capacity to meet expanding demand with consequent loss of market share to competitors. Since 1973, more sluggish levels of demand, unemployment, and slacker use of capacity have made entrepreneurs less euphoric.

THE GROWTH ACCOUNTS

In attempting to explain the complex developments which have occurred since 1950 in terms of detailed growth accounts, the following analysis is concentrated on the experience of six countries, i.e. the three successive lead countries in the capitalist epoch — the Netherlands, UK, and USA; the follower country with the lowest starting-point and fastest growth — Japan; and the two biggest West European countries — France and Germany. This sample gives a good idea of the full range of experience, though the average degree of acceleration and slow-down for the six was bigger than for the sixteen.

The growth accounts go successively through the main features which could have significant explanatory value, starting with changes in the quantity of labour and capital inputs, and changes in their quality. Changes in the momentum of labour input explain only part of the post-war acceleration and slow-down. The movement in capital inputs is much more closely correlated to those for output. Nevertheless, as capital gets a much lower weight in our accounts than labour, the movement of joint factor input still leaves a good deal unexplained. To reach more satisfactory levels of explanation, it is necessary to go beyond factor inputs and look at other special features. The three most important were the opening of the economies to trade, gains from structural change, and accelerated technological diffusion. Three smaller influences were economies of scale, energy substitution which was a drag on productivity after 1973, and the impact of natural resource discoveries in the Netherlands and the UK.

Growth accounting of this type cannot provide a full causal story.[2] It deals with 'proximate' rather than 'ultimate' causality and registers the facts about growth components; it does not explain the elements of policy or circumstance, national or international, that underlie them (and which are explained in Chapter 6 below), but it does identify which facts need more ultimate explanation. This kind of exercise forces one to merge and match data in a way that provides valuable cross-checks on the consistency and plausibility of the basic growth indicators both for individual countries and across countries. It is particularly useful in avoiding or identifying double counting and in tracking down the complex interactions among causal components.

The transparency of the operation is a major advantage, for although on significant points there are large judgemental elements, and the quality of the evidence ranges from hard facts to hunches, the individual steps in the analysis are clearly identified and the reader has building bricks for alternative hypotheses. He or she can augment, truncate, or reweight to taste, provided

TABLE 5.3. *Growth of Labour Potential and Input, 1913–1987* (annual average compound growth rates)

	France			Germany			Japan		
	1913–50	1950–73	1973–87	1913–50	1950–73	1973–87	1913–50	1950–73	1973–87
Labour force	0.04	0.46	0.62	0.73	0.74	0.55	0.88	1.67	0.95
Employment	0.04	0.43	0.02	0.55	1.08	0.00	0.89	1.70	0.84
Labour hoarding	0.00	0.00	0.00	0.00	0.00	0.00	−0.78	1.26	−0.22
Hours per person	−0.80	−0.36	−0.98	−0.30	−1.08	−0.77	−0.48	−0.15	−0.25
Quantity of labour	−0.76	0.06	−0.96	0.24	−0.02	−0.77	−0.37	2.83	0.36
Impact of education	0.52	0.56	0.80	0.35	0.27	0.15	0.87	0.74	0.64
Sex-mix	0.00	−0.36	−0.19	−0.05	−0.04	−0.08	0.01	0.00	−0.05
Quality of labour	0.52	0.20	0.61	0.30	0.23	0.07	0.88	0.74	0.59
Augmented labour input	−0.24	0.26	−0.35	0.54	0.21	−0.70	0.51	3.59	0.95

	Netherlands			UK			USA		
Labour force	1.59	0.98	1.61	0.52	0.48	0.63	1.29	1.51	2.03
Employment	1.55	0.97	1.09	0.51	0.49	0.00	1.29	1.51	1.93
Labour hoarding	0.00	0.00	0.00	0.00	0.00	0.00	0.00	0.00	0.00
Hours per person	−0.45	−1.00	−1.65	−0.79	−0.64	−0.58	−0.90	−0.36	−0.47
Quantity of labour	1.10	−0.04	−0.58	−0.28	−0.15	−0.58	0.35	1.16	1.45
Impact of education	0.38	0.62	0.80	0.47	0.29	0.47	0.59	0.68	0.63
Sex-mix	0.01	−0.06	−0.35	−0.02	−0.12	−0.16	−0.09	−0.18	−0.21
Quality of labour	0.39	0.56	0.45	0.45	0.17	0.31	0.50	0.50	0.42
Augmented labour input	1.49	0.52	0.13	0.17	0.02	−0.27	0.85	1.67	1.87

Sources: First row from Table C.7. Second row from Table C.8. Third row applies only to Japan — see text. Fourth row from Table C.9. Fifth row results from multiplication of row 2, 3, and 4 effects. Sixth row from sources cited in Table 3.8, giving a weight of 1 to primary education, 1.4 to secondary, and 2 to higher, with increments weighted by 0.6. Seventh row from Table C.2 with weight of 0.6 for females and 1.0 for males. Eighth row results from multiplication of row 6 and 7 effects. Ninth row results from multiplication of row 5 and 8 effects.

that the 'explanations' of growth are logical, consistent, and explicit. This is why the present chapter sticks to simple, standardized, and rather conservative hypotheses throughout, so that the impact of a change in assumptions for a particular growth component can be readily traced on a cross-country basis.

THE QUANTITY AND QUALITY OF LABOUR INPUT

Table 5.3 shows detailed estimates of the quantity and quality of labour input for our six sample countries. In France, Japan, and the USA one can observe an acceleration in labour input in 1950–73 which is of some help in explaining output acceleration, but in Germany, the Netherlands, and the UK, total augmented labour input actually increased more slowly in the golden age. However, after 1973, in all countries except the USA, labour input grew more slowly or declined and had some explanatory force in accounting for the slow-down.

Unemployment

The first row in Table 5.3 shows growth in labour force. Until 1973–87, there were no big differences in the growth of labour force and employment, except in Germany (1950–73), where employment rose faster because there was still significant labour slack, which was not fully absorbed until 1960.

In 1973–87, employment rose less than the labour force in all the countries, with the difference being particularly marked in the European countries. There clearly was some loss of potential output through unemployment. Increased labour slack was consciously accepted by most governments as a means of checking the inflation which occurred after 1973.

Labour Hoarding

In normal circumstances one would not expect labour hoarding or dishoarding to be significant in advanced capitalist economies over a period of years because market forces cause workers to be laid off if they cease to be productive. Even in cases where labour legislation restricts the freedom to fire and hire, labour turnover will permit substantial attrition over a number of years.

However, Japan differs from the other countries in two respects. First, a very high ratio of employed persons are self-employed or family workers. In 1950 this proportion was approximately 60 per cent and in 1973 it was 31 per cent; in the USA, by contrast, the corresponding shares were 20 and 10 per cent. Second, a significant portion of wage and salary earners have lifetime job security, which employers can guarantee because wages are very flexible, with mid- and end-of-year bonuses that move with business profits and can amount to a third of earnings in normal times. These bonuses can be squeezed to zero in depressed conditions, which makes it fairly easy for employers to retain workers rather than lay them off. In these conditions, one cannot rely on Japanese unemployment figures to provide a guide to labour slack.

In 1950, it is clear that there was a large degree of under-employment in Japan. Official unemployment stood at only 1.9 per cent of the labour force, but GDP was only three-quarters of its 1941 level. I assumed that labour hoarding amounted to a quarter of employment in 1950 and that this labour had been fully dishoarded by 1973. I assumed that there was also a modest 3 per cent labour hoarding in the slow-down of growth after 1973. The results of these hypotheses can be seen in Table 5.3, where labour hoarding had a significant explanatory impact in 1913–50, and dishoarding favoured growth in 1950–73.

Hours per Person

Over the long run, working hours per person have fallen substantially in these countries. In fact they are now about half of their level in 1870. Not all of the increase in productivity has been taken out in real income; people have preferred additional leisure in the form of longer vacations and weekends, cutting daily hours substantially, and taking sickness absence more freely. These processes were operative in all three of the periods considered in Table 5.3, and they differ in pattern from country to country. In Germany and the Netherlands the reduction of hours was faster in 1950–73 than in 1913–50, so this does not help explain growth acceleration. In France, Japan, the UK, and the USA the rate of reduction was smaller in 1950–73 and thereby contributed modestly to explain output acceleration. In 1973–87 there were significant cuts in French and Dutch working hours as a result of deliberate government policy to mitigate unemployment by job-sharing schemes.

Educational Level

The sixth and seventh rows in the table represent the effect of two types of quality changes in labour input. The first and most important of these is the rise in levels of education. From 1913 to 1987 the average level of education rose from about 6.4 to more than 11 years, and the increase was continuous. The increase since 1913 has been concentrated on secondary and higher education, where the impact on human capital is greater than at primary level, if one judges by the relative earnings of those with different types of education. The stock of higher education rose eightfold from 0.1 to about 0.8 of a year from 1913 to 1987 (average for the six countries), and the stock of secondary education rose almost threefold from 1.7 to 4.8 years (see Table 3.8).

The rate of improvement in educational levels was generally no more important in the golden age than it was in 1913–50, nor was there any general slow-down after 1973, so this labour quality factor is not important in explaining growth and slow-down.

It is generally accepted that the strong positive relationship between wage levels and levels of education is partly due to the fact that levels of education are correlated with intelligence, family connections, or the impact of credentialist practices, so some compression of educational weights is warranted.[3]

Sex-Mix

The second quality adjustment is meant to correct for changes in the sex composition of labour. Female labour compensation rates are almost universally lower than those for males, partly because more women than men work on a temporary or part-time basis, or because their processes of skill acquisition are less favourable than for males because of short or interrupted labour market participation (e.g. exits for childbearing). Some of the lower pay for women is, of course, due to discrimination and to this extent earnings differentials exaggerate the difference in the 'quality' of male and female labour. In this exercise, female labour is given a weight of 0.6 and males a weight of 1.0, which is intended to be a rough average of the female/male wage differential. When the proportion of females in the labour force increases (see Appendix Table C.2), which has been the overwhelming long-run trend, the sex-mix adjustment will be negative.

In the golden age, changes in the sex-mix are of no help in explaining the growth acceleration in five of the six countries, and they play only a modest role in explaining the growth slow-down.

Augmented Labour Input

In the last row of Table 5.3, the impact of all the quantitative and qualitative changes is combined into a measure of 'augmented' total labour input. We can see a significant acceleration in the total measure for France, Japan, and the USA for the golden age, but not for the other three countries. For the period since 1973, augmented labour input decelerated in all the countries except the United States.

THE QUANTITY AND QUALITY OF CAPITAL INPUT

The movement of capital stock provides a much more powerful degree of explanation for growth acceleration and slow-down than we got from the analysis of labour input.

Non-Residential Capital

Tables 5.4 and 5.5 show the growth of non-residential gross and net capital stocks respectively since 1913. The figures exclude housing, inventories, and land. The problem of capacity use is ignored; it is simply assumed in the following growth accounts that capital input moved parallel with changes in the stock of assets.

It emerges very clearly from Tables 5.4 and 5.5 that 1950–73 was a period of supergrowth for productive capital. The average growth of gross capital stock was 5.8 per cent a year — about three times the pace achieved in 1913–50. For net stock the acceleration was sharper — an average 6.8 per cent a year compared with 1.9 per cent. The acceleration in average growth of capital per person employed was even greater, 4.8 per cent a year compared with an average of 1.2 per cent for 1913–50 for gross stock and 5.7 compared with 1.1 for net stock. The 1950–73 acceleration affected all countries, but it was much greater in

TABLE 5.4. *Rate of Growth of Gross Non-Residential Capital Stock, 1890–1987* (annual average compound growth rates)

	1890–1913	1913–50	1950–73	1973–87
France	n.a.	(1.21)	5.12	4.49
Germany	(3.09)	(1.06)	6.60	3.45
Japan	3.00	3.85	9.08	7.57
Netherlands	n.a.	(2.43)	5.83	3.33
UK	2.02	1.47	5.11	2.87
USA	5.42	2.09	3.24	3.28
Arithmetic average	3.38	2.02	5.83	4.17

Sources: The figures are the total for non-residential structures plus machinery and equipment. Standardized estimates from Appendix D, except for the figures in brackets, which are proxies. France 1913–50 is based on linkage of two rough point estimates cited by J. J. Carré, P. Dubois, and E. Malinvaud, *La Croissance française*, Seuil, Paris, 1972, p. 204. Germany 1913–50 is from estimates of Hoffmann, Grumbach, and Hesse, which are based on a mix of valuation and perpetual inventory techniques as described in A. Maddison, *Phases of Capitalist Development*, pp. 225–6. For the Netherlands, it was simply assumed that the 1913–50 capital stock movement paralleled the movement in GDP. My perpetual inventory estimates of Japanese stock of non-residential structures extend back only to 1925. For 1913–25 I used the gross stock movement shown in K. Ohkawa and M. Shinohara, *Patterns of Japanese Economic Development*, Yale, 1979, pp. 366–7, described on their pp. 189–93.

the follower countries, which were catching up with US levels of productivity and technology, than it was in the USA.

In 1973–87, gross capital stock growth decelerated notably in all the follower countries, but not in the USA. The deceleration was even more marked for net stock and applied to all the countries. In spite of the deceleration, the growth of capital stock after 1973 was generally higher that it was before 1950.

Over the very long term, gross capital stock (i.e. cumulated gross investment minus scrapping) and net capital stock (i.e. cumulated gross investment minus depreciation) indices tend to move in a broadly similar way (though the level of the net stock is

TABLE 5.5. *Rate of Growth of Net Non-Residential Capital Stock, 1913–1987* (annual average compound growth rates)

	1913–50	1950–73	1973–87
France	(1.21)	6.37	3.69
Germany	(1.06)	7.71	2.68
Japan	3.59	10.21	6.65
Netherlands	(2.43)	6.89	2.16
UK	1.55	5.70	2.25
USA	1.69	3.84	2.59
Arithmetic average	1.92	6.79	3.34

Sources: As for Table 5.4.

smaller), but when capital formation (investment) is accelerating, as in 1950–73, the net stock grows faster than the gross. When capital formation decelerates, as in 1973–87, net stock grows more slowly than gross.

The movement in gross capital stock gives a somewhat exaggerated representation of changes in capital's ability to contribute to production. All assets are treated as if they maintained their initial value in real terms until they are scrapped. Thus, an automobile is counted at its purchase value in the gross stock until it is scrapped. The car will indeed perform its physical function to the end but its reliability will be interrupted by increasingly frequent breaks for repairs. By contrast, in the net stock, the car is written down steadily throughout its life. The net valuation thus represents the discounted value of the car to its owner. However, the decline in its second-hand value overstates the decline in its functional attributes. The car's contribution to current production is much closer to the gross than to the net valuation. I have therefore used the gross stock to measure capital input.

Embodied Technical Progress, the Age Effect, and the Quality of Capital

If technical progress had come to an end and the follower countries had caught up with the leader, there would not be

TABLE 5.6. *Rate of Growth of Gross Non-Residential Capital Stock per Person Employed, 1890–1987* (annual average compound growth rates)

	1890–1913	1913–50	1950–73	1973–87
France	n.a.	1.18	4.78	4.54
Germany	1.48	0.51	5.47	3.45
Japan	1.94	2.94	7.26	6.38
Netherlands	n.a.	0.88	4.81	2.22
UK	1.08	0.96	4.59	2.87
USA	3.41	0.83	1.69	1.33
Arithmetic average	1.98	1.22	4.77	3.47

Sources: Capital stock as for Table 5.4, employment from Appendix C.

TABLE 5.7. *Rate of Growth of Net Non-Residential Capital Stock per Person Employed, 1913–1987* (annual average compound growth rates)

	1913–50	1950–73	1973–87
France	1.18	5.92	3.67
Germany	0.51	6.57	2.68
Japan	2.68	8.37	5.76
Netherlands	0.88	5.86	1.06
UK	1.04	5.18	2.27
USA	0.42	2.27	0.65
Arithmetic average	1.12	5.70	2.68

Source: As for Table 5.6.

profitable scope for capital deepening, i.e. the increases in capital per employee, which we see in Tables 5.6 and 5.7.

The increase in capital per employee is made profitable by technical progress. If progress is steady, each successive year will bring an equi-proportional vintage bonus to the capital stock, because better technology is embodied in it. Schumpeter thought that technical progress came in big spurts, but I think he was wrong and that there are stronger grounds for assuming its impact to be relatively steady.[4]

Given the complexity of forces affecting economic growth, it is not easy to infer at all exactly what this rate of progress has been. The figure of 1.5 per cent per annum for embodied technical progress which is used here is intended to be illustrative, not definitive.[5] In any case, it reflects only part of the influence of technical progress. Some of it is disembodied and affects the size of the residual in our growth accounts.

Even if a steady rate of technical progress is embodied in the flow of new investment (as I assume here), its impact on the capital stock will be uneven if the rate of capital formation is changing over time. The investment boom of the golden age brought a reduction in the average age of capital stock even though life expectation of the two major categories of assets was unchanged. This is the same effect one can observe with demographic phenomena. If life expectation is unchanged and there is a prolonged increase in fertility, there is an increase in the proportion of younger people and the average age of the population drops.

Table 5.8 shows that there was a big reduction in the average

TABLE 5.8. *Average Age of Gross Non-Residential Capital Stock, 1890–1987* (years)

	1890	1913	1950	1973	1987
France	n.a.	n.a.	17.89	9.88	12.26
Germany	n.a.	n.a.	16.75	10.29	13.40
Japan	n.a.	11.00	13.00	7.44	11.05
Netherlands	n.a.	n.a.	17.76	10.74	14.30
UK	16.50	15.50	12.74	9.97	12.25
USA	12.80	13.67	16.30	12.65	13.96

Source: Derived from 'standardized' capital stock estimates in Appendix D. Estimates were broken down by non-residential structures with life expectation of 40 years, and machinery and equipment with life expectation of 15 years.

age of the capital stock from 1950 to 1973. The impact of this 'age effect' was to raise the average gain for the six countries from the embodiment effect in 1950–73 to 1.85 per cent a year instead of the 1.5 per cent a year that would have prevailed if the average age had not been reduced. This happened because a lowering of the average age pushes the average level of technical performance embodied in the stock nearer to the 'best practice' technology embodied in the newest capital.[6] When the average age of capital falls, the gap between average and best practice is reduced; and, conversely, when the average age rises, the gap widens.

Residential Capital

Although residential capital represents a sizeable share in the national wealth (e.g. residential assets in the USA in 1987 were about 36 per cent of the combined value of residential and non-residential fixed capital — see Tables 3.10 and 3.12 above), it plays a more modest role in our growth accounts than does non-residential capital. There are two main reasons for this. In the first place, residential capital plays a negligible role in transmitting new technology into the production process, so I assume there is no embodied quality effect as there was for non-residential capital. Secondly, I give residential capital a lower weight than its asset value might suggest. In the following accounts capital input is given a weight of 0.3, with 0.23 for non-residential and 0.07 for residential capital, i.e. residential gets about a quarter of the weight for total capital input and non-residential capital gets about three-quarters of the weight.

In some growth-accounting exercises, the individual items in the capital stock are weighted at their asset values, and it is assumed that the stock thus calculated provides a proxy measure of the potential service flow derivable from the assets. However, the service flow is better approximated by using rental rather than asset weights. As housing has a much longer life than other capital assets, its rental weight is a good deal lower than its asset weight, e.g. a $100,000 house has an asset value ten times as big as a $10,000 car, but the annual rental value of the house might be as low as 5 per cent, whereas for the car it might be 30 per cent (as the annual car rental must cover a much bigger element of depreciation than in the case of the house). Thus the rental

TABLE 5.9. *Rate of Growth of Gross Residential Capital Stock, 1913–1987* (annual average compound growth rates)

	1913–50	1950–73	1973–87
France	0.39	2.80	2.27
Germany	0.40	5.02	3.08
Japan	0.12	6.15	5.16
Netherlands	1.94	2.63	2.64
UK	1.65	2.94	2.56
USA	1.82	3.29	2.48

Source: See note to Table 3.12.

weights would be in the ratio of 5 : 3 compared with 10 : 1 as for the asset ratio.

TOTAL FACTOR INPUT

In Table 5.10 the quantity and quality changes in capital stock are combined in a measure of augmented capital input. This is then combined with the augmented labour input we already derived in Table 5.3, to produce augmented joint factor input.

The weights used for combining inputs of labour and capital are 0.7 for labour and 0.3 for capital, i.e. they correspond roughly to the share of labour and capital income in GDP. This is the normal way in which factor inputs are weighted together in growth accounts.[7]

FORCES SUPPLEMENTAL TO FACTOR INPUT

The augmented joint factor input measure in Table 5.10 explains only part of the growth acceleration of the golden age. It is therefore necessary to look at other forces favouring the process of post-war growth acceleration. If such forces are to be a legitimate element of explanation, they must be allowed to exercise symmetric types of influence in explaining slow-down as well.

TABLE 5.10. *Growth of Capital Inputs, and Augmented Joint Factor Input, 1913–1987 (annual average compound growth rates)*

	France			Germany			Japan		
	1913–50	1950–73	1973–87	1913–50	1950–73	1973–87	1913–50	1950–73	1973–87
Quantity of non-residential capital	1.21	5.12	4.49	1.06	6.60	3.45	3.85	9.08	7.57
Quality of non-residential capital	1.50	2.10	1.34	1.50	1.89	1.24	1.43	1.81	1.22
Augmented non-residential capital input	2.73	7.33	5.90	2.58	8.61	4.73	5.34	11.05	8.58
Residential capital	0.39	2.80	2.27	0.40	5.02	3.08	0.12	6.15	5.16
Augmented labour input	−0.24	0.26	−0.35	0.54	0.21	−0.70	0.51	3.59	0.95
Augmented joint factor input	0.48	2.02	1.24	1.00	2.42	0.79	1.57	5.44	2.95

	Netherlands			UK			USA		
	1913–50	1950–73	1973–87	1913–50	1950–73	1973–87	1913–50	1950–73	1973–87
Quantity of non-residential capital	2.43	5.83	3.33	1.47	5.11	2.87	2.09	3.24	3.28
Quality of non-residential capital	1.50	1.94	1.15	1.61	1.67	1.27	1.41	1.71	1.36
Augmented non-residential capital input	3.97	7.88	4.52	3.11	6.86	4.18	3.53	5.01	4.69
Residential capital	1.94	2.63	2.64	1.65	2.94	2.56	1.82	3.29	2.48
Augmented labour input	1.49	0.52	0.13	0.17	0.02	−0.27	0.85	1.67	1.87
Augmented joint factor input	2.09	2.32	1.30	0.94	1.76	0.93	1.53	2.54	2.55

Sources: Derived from Tables 5.3, 3.11 and Appendix D. Augmented non-residential capital input is derived by multiplying the impact of the increments in quantity and quality. In combining capital and labour input to derive joint factor input, the labour items are given a weight of 0.7, non-residential capital a weight of 0.23, and residential capital a weight of 0.07.

There are four general forces which seem sufficiently powerful to be worth explicit consideration and which affected the efficiency of resource allocation. The first is the vast opening up of these economies to international trade which occurred as a result of policy action in the post-war period. The second is the influence of structural change which pulled labour out of low performance or less dynamic sectors towards more dynamic parts of the economy. The third is the accelerated diffusion of technology, and the fourth is economies of scale.

A fifth effect, of generally smaller significance and affecting mainly the deceleration phenomenon, is that which derived from the OPEC shocks, which induced an unusually sharp process of energy economy which had some costs for growth. In the context of energy we also consider the effect of natural resource discoveries of Groningen gas in the Netherlands and North Sea oil and gas in the UK.

Impact of Foreign Trade

It is clear from Table 5.11 that there was a huge acceleration in the growth of foreign trade in all these countries in 1950–73. It affected all six countries. It was modest in the USA, and greatest in Germany and Japan. Average trade volume (exports plus imports) rose by 9.4 per cent a year for 23 years compared with a growth of less than half a per cent a year from 1913 to 1950.

After 1973 trade growth slowed down everywhere and was more even across the whole range of countries, so it obviously played a role in explaining deceleration, though trade growth after 1973 was still high by pre-1950 standards.

I have assumed that the expansion of foreign trade produced gains from economies of scale and specialization amounting to 15 per cent in each of the periods.

The importance of the trade effect shown in row 3 of Table 5.19 depends on the size of trade relative to GDP in Table 5.12 as well as the growth rates shown in Table 5.11.

These changes in the role of trade had a good deal to do with policy. In the 1913–50 period, trade was adversely affected by the growth of tariffs, quantitative restrictions, and exchange controls. In 1950–73 tariffs were removed on a great part of the inter-trade of the four European countries within the European

TABLE 5.11. *Rate of Growth in Volume of Foreign Trade, 1913–1987* (annual average compound growth rates)

	1913–50	1950–73	1973–87
France	0.69	8.79	4.11
Germany	−2.36	11.95	3.79
Japan	1.06	15.54	6.60
Netherlands	1.00	9.26	2.98
UK	0.20	4.50	3.48
USA	2.35	6.47	4.08
Arithmetic average	0.49	9.42	4.17

Sources: Volume of exports derived from Appendix F. Volume of imports 1913–50 from A. Maddison, 'Growth and Fluctuation in the World Economy, 1870–1960', *Banca Nazionale del Lavoro Quarterly Review*, June 1962, except Japan, which is from K. Ohkawa and H. Rosovsky, *Japanese Economic Growth*, Stanford, 1973, p. 302. 1950 onwards from IMF, *International Financial Statistics*, and OECD, *Economic Outlook*. The figures in this table are simply the average of the growth rates for export and import volume.

Community, and on a world-wide basis for all six countries by successive GATT rounds. Quantitative restrictions on non-agricultural products were greatly reduced, and exchange controls were reduced to negligible proportions. After 1973, the tariff reduction process was less significant, and little progress was made in reducing barriers to agricultural trade.

The post-1950 growth in trade strengthened international specialization and competition and gave European countries some of the traditional American advantages of large internal markets. The prices of products entering foreign trade, both imports and exports, rose a good deal less than the overall price level.

Impact of Structural Change

Economic growth has been accompanied by massive changes in economic structure, whose long-run pattern has been similar in

TABLE 5.12. *Ratio of Commodity Exports plus Imports to GDP at Current Market Prices, 1913–1987* (%)

	1913	1950	1973	1987
France	30.9	21.4	29.2	34.8
Germany	36.1	20.1	35.3	46.8
Japan	30.1	16.4	18.2	16.1
Netherlands	100.0	70.9	74.8	86.5
UK	47.2	37.1	37.6	42.0
USA	11.2	6.9	10.8	15.2
Arithmetic average	42.6	28.8	34.3	40.2

Sources: Post-war from IMF and OECD. Pre-war from national sources.

all the advanced economies. Employment in agriculture declined from an average of 34 per cent in these six countries in 1913 to 5 per cent in 1987. The service share rose from 33 per cent in 1913 to 64 per cent. Industry accounted for 32 per cent of employment in 1913, rose to a peak of 39 per cent in 1960, and has now fallen back below its 1913 share. The same bell-shaped pattern of relative industrialization and de-industrialization has affected all of the countries, though its timing has varied, peaking earliest in the Netherlands, UK, and USA and latest in Japan.

Structural changes reflect two basic forces that have operated on all the countries as they reached successively higher levels of real income and productivity. The first of these is the elasticity of demand for particular products, which has been rather similar across countries at given levels of income (particularly as relative price structure has moved in a similar direction). These demand forces have reduced the share of agricultural products in consumption and have raised the share of industry and services. The second basic force has been the differential pace of technological advance between sectors. Productivity growth has been slower in services than in commodity production; partly because of the intrinsic character of many personal services; partly because of measurement conventions that exclude the possibility of productivity growth in some services.

Historically, agriculture has had low productivity levels, and

Table 5.13. *Sectoral Labour Productivity Growth (Gross Value Added per Person Employed), 1913–1987* (annual average compound growth rate)

	1913–50	1950–73	1973–87
Agriculture, Forestry, and Fishing			
France	1.4	5.9	4.8
Germany	−0.4	6.3	3.7
Japan	0.3	7.3	1.9
Netherlands	n.a.	6.0	4.4
UK	2.5	4.6	3.6
USA	1.6[a]	5.4	2.4
Average	1.1	5.9	3.5
Industry (including Construction)			
France	1.5	5.2	3.9
Germany	1.3	5.6	2.0
Japan	2.4	9.5	3.9
Netherlands	n.a.	5.6	1.5
UK	1.4	2.9	2.8
USA	1.5[a]	2.2	1.1
Average	1.6	5.2	2.5
Services			
France	0.4	3.0	0.8
Germany	−0.2	2.8	2.0
Japan	0.3	4.0	2.2
Netherlands	n.a.	1.8	0.4
UK	0.7	2.0	1.1
USA	1.0[a]	1.4	0.2
Average	0.4	2.5	1.1

[a] 1909–48.

Sources: GDP by sector from sources described in Appendix A. Employment derived from Appendix C.

TABLE 5.14. *Level of Output per Person Employed by Sector, 1987* (% of average for whole economy)

	Agriculture	Industry	Services
France	68	122	93
Germany	42	100	105
Japan	34	125	95
Netherlands	97	121	95
UK	86	131	87
USA	92	121	93
Average	70	120	95

Sources: Sectoral GDP at factor cost in national prices from *National Accounts of OECD Countries* (without deduction of inputed bank service charges). Employment from Appendix C, Tables C.5 and C.8.

the exodus from this sector to higher productivity employment in industry and services was an important element in structural 'progress'. After the Second World War, high levels of overall demand speeded up this transfer, sucking labour out of agriculture at a faster pace. Helped by technical advances, agricultural productivity speeded up in the post-war period and has in most cases been faster than in industry (see Table 5.13). Because the relative level of agricultural productivity is higher now, and farm employment is much smaller, the scope for this kind of structural gain has been eroded.

In the past there was always some structural 'drag' from movement into services, where productivity grew slowly. This was compensated by the relatively high level of service productivity, but the service productivity level has now fallen below that in industry in all countries except Germany (see Table 5.14).

The contribution of structural change to growth was generally most favourable in 1950–73, and least in 1973–87 (when it was actually negative for four of the countries — see Table 5.19). Over the whole period since 1913, it was most favourable for Japan, which started with the biggest agricultural sector, and least favourable for the Netherlands and UK, which had the most 'mature' economic structures.

In Table 5.19 the structural effect is calculated by comparing the actual growth rate of output per man and what it would have been if the structure of employment had not changed and if productivity growth within each sector had remained as actually experienced.

Accelerated Diffusion of Technology

The technology developed at the frontier (mainly by the lead country, the USA) is not a set of blueprints which can easily be copied by the follower countries. Its assimilation and exploitation involve a good deal of adaptive innovation, learning by doing, and effective mechanisms for the transfer of know-how.[8]

The 1913–50 period was not auspicious for such transfer. Travel, communication, study abroad, and exchange of ideas was blocked by two world wars, and was not helped by the depression and autarkic policies of the 1930s. After the war, the possibilities for technology transfer between these countries were greatly increased. The Marshall Aid programme from 1948 to 1952 involved a major effort by governments which was continued for a decade thereafter by the creation of productivity agencies which sent teams of experts and industrialists to the USA to help narrow the technological gap. The stock of US private direct

TABLE 5.15. *Research and Development Expenditure per Person Employed, 1960–1987* ($ in 1985 US relative prices)

	1960	1973	1987
France	207	448	761
Germany	179	475	848
Japan	89	342	757
Netherlands	291	514	687
UK	343	480	650
USA	809	814	1,074

Source: C. Freeman and A. Young, *The Research and Development Effort*, OECD, Paris, 1965, and subsequent survey information supplied by the OECD Science and Technology Directorate. OECD proportions were applied to the GDP figures in Appendix A. Employment from Appendix C.

investment in Europe (in real terms, using the GDP deflator) rose by 11.1 per cent a year from 1950 to 1973, and from 1973 to 1987 by 3.3 per cent a year. From 1929 to 1950, by contrast, it fell by an amount equivalent to 1.2 per cent a year.

Research and development activity in the follower countries is as important in absorbing and developing existing best practice from the lead country as it is in developing completely new ideas. It is not possible to quantify these efforts in a comparative way before the early 1960s, but the convergence in the scope of the R. & D. effort is clear from Table 5.15.

Given the difficulty of obtaining a direct aggregate measure of the multifarious efforts to accelerate technological diffusion and their impact, I have, as a very crude proxy, assumed that such activities explain 10 per cent of the convergence in labour productivity after 1950 (see Table 5.2). For 1913–50 I assumed that diffusion activities were at a standstill.

Economies of Scale

In strict neo-classic versions of growth accounts it is assumed that there are constant returns to scale, but in all of Denison's studies a significant scale bonus is assumed to occur as national markets increase in size. Other analysts give scale economies much bigger emphasis than Denison, e.g. Adam Smith, Nicholas Kaldor, or new growth theorists such as Paul Romer,[9] but empirical evidence on economies of scale is scarce. They are difficult to calculate in cross-section studies because of inter-firm variations in vintages of capital. In time series, it is difficult to disentangle scale economies from other manifestations of technical progress, learning by doing, etc. There is therefore considerable room for dispute about their importance. In my view the most significant economies of scale are those we already captured in activities whose products enter international trade, which is the main reason small countries are able to profit by them. I therefore assume that the scope for such economies within domestic markets is relatively modest — 3 per cent of GDP growth.

Impact of the Energy Crisis

Table 5.16 shows the total energy requirements (in terms of oil equivalent) of these countries for the bench-mark years from 1913

TABLE 5.16. *Energy Consumption, 1913–1987* (million tons of oil equivalent)

	1913	1950	1973	1987
France	55.7	64.6	179.7	202.9
Germany	93.8	93.9	266.2	273.3
Japan	21.7	44.9	338.9	378.4
Netherlands	7.8	14.2	63.2	72.2
UK	157.5	158.3	220.8	210.5
USA	436.5	866.0	1,759.4	1,857.6

Sources: 1913 and 1950 from A. Maddison, *Journal of Economic Literature*, June 1987, p. 693. 1973 from *Energy Balances of OECD Countries, 1970/85*, IEA (International Energy Agency), Paris, 1987. 1987 supplied by IEA. The figures are adjusted for territorial change.

TABLE 5.17. *Rate of Growth of Energy Inputs, 1913–1987* (annual average compound growth rates)

	1913–50	1950–73	1973–87
France	0.40	4.50	0.87
Germany	0.00	4.63	0.19
Japan	1.98	9.19	0.79
Netherlands	1.64	6.71	0.96
UK	0.01	1.46	−0.34
USA	1.87	3.13	0.39
Arithmetic average	0.98	4.94	0.48

Source: Derived from Table 5.16.

to 1987. Clearly, economic growth required a very large increase in energy consumption over the long run. However, the pace of growth in energy demand has been uneven. In the golden age, 1950–73, inputs of energy accelerated greatly. They slowed down sharply after 1973 in response to slower economic growth and the reinforcement of energy economy which rising prices induced.

Average oil import prices rose from $19.54 a ton in 1973 to $243.53 in 1980, and then fell to $132.46 in 1987.[10]

Over the long term, all these countries have been engaged in a successful process of energy economy. It can be seen from Table 5.18 that the process of energy economy applied in 17 out of 18 country cases shown in the first three columns. The only exception was the Netherlands in 1950–73, when the economy became more energy intensive under the influence of the large discoveries of cheap natural gas in Groningen.

From the fourth column of Table 5.18 it can be seen that in 1950–73 there was a relaxation of the process of energy economy from the situation that had applied in 1913–50. This relaxation was due primarily to the arrival of massive new sources of cheap energy from the Middle East.

In 1973–87, by contrast, the OPEC shocks raised the price of imported oil more than twelvefold in two successive leaps, and this led to a sharp reinforcement of the process of energy economy. In all the countries, this slowed down economic growth because of the costs involved in cutting the energy input ratio sharply and substituting between different kinds of fuel or sources of supply.

I used shares of energy input in GDP at factor cost as weights to calculate the impact of changes in the rate of energy economy on growth. For 1973–87 the weight of energy was bigger and so was the pace of economization, so all the countries show a negative result in Table 5.19 row 6. For 1950–73 the impact was negligible except in the Netherlands.

The energy effect thus derived is intended to reflect only the direct impact on GDP growth. The OPEC shocks also had indirect effects on the conjuncture as they exacerbated inflation and created payments problems that induced deflationary policies. They also worsened real GNP (as opposed to real GDP) through their terms of trade effect.

Effects of Natural Resource Discovery

The foregoing analysis ignored the role of natural resources. They normally play no role in growth acceleration or deceleration, and in advanced economies they have rather a low weight.

However, natural gas discoveries made a significant contribution to boosting Dutch output from 1968 to 1974, and North Sea

TABLE 5.18. *Changes in Energy Input Ratios, Energy Economy Relaxation, and Reinforcement, 1913–1987* (annual average compound growth rates)

	Change in Energy/GDP Ratio			Energy economy relaxation 1950–73	Energy economy reinforcement 1973–87
	1913–50	1950–73	1973–87		
France	−0.74	−0.51	−1.24	0.23	−0.73
Germany	−1.26	−1.12	−1.52	0.14	−0.40
Japan	−0.25	−0.07	−2.83	0.18	−2.76
Netherlands	−0.73	1.88	−0.81	2.61	−2.69
UK	−1.24	−1.52	−2.18	−0.28	−0.66
USA	−0.90	−0.50	−2.17	0.40	−1.67
Arithmetic average	−0.85	−0.31	−1.79	0.54	−1.48

Source: First three columns derived by dividing rate of change in energy input by rate of change in GDP. Fourth column shows the relaxation of energy saving as reflected by the difference between columns 1 and 2. Fifth column shows reinforcement of energy saving, as reflected by the difference between columns 2 and 3.

oil and gas discoveries helped British growth from 1976 to 1983. Neither of these bonanzas had counterparts in the other four countries under consideration (but they did affect Norway). For the period 1950–73 as a whole, natural gas growth accounted for 8 per cent of the large increment in Dutch real GDP, and increases in North Sea output represented 15 per cent of the rather small 1973–87 increase in British GDP. If all of this new energy pro- duction had been a natural resource windfall, it would have raised the Dutch growth rate by 0.24 percentage points a year for 1950–73, and British growth by 0.28 points a year for 1973–87.

However, North Sea oil and gas exploitation involved heavy capital costs, so we treat only half of its contribution as a windfall. In the Netherlands, the costs of developing Groningen gas were a good deal smaller, so 80 per cent of the increment was taken to be a windfall.

EXTENT TO WHICH GROWTH IS EXPLAINED

We are now in a position to judge the cumulative impact of the forces we have identified and quantified, i.e. the two main factor inputs, adjusted for quality change, and the six supplemental influences discussed above.

Table 5.19 shows their cumulative impact in explaining growth and the size of the unexplained residual. Table 5.20 shows the percentage of growth which has been explained in each period, and the average degree of explanation for 1913–87 as a whole. The average level of explanation is over three-quarters and is rather stable over the three periods. The least satisfactory ex- planation is that for French growth in 1913–50.

EXTENT TO WHICH ACCELERATION AND DECELERATION ARE EXPLAINED

Table 5.21 shows the degree to which the acceleration of 1950–73 is explained. The average degree of explanation is about three- quarters. The lowest degree of explanation is for Germany and there is mild over-explanation of the US acceleration.

The average level of explanation for the slow-down for the six

TABLE 5.19. *Identifiable Forces Explaining GDP Growth, 1913–1987* (annual average compound % contributions to GDP growth)

	France			Germany			Japan		
	1913–50	1950–73	1973–87	1913–50	1950–73	1973–87	1913–50	1950–73	1973–87
GDP	1.15	5.04	2.16	1.28	5.92	1.80	2.24	9.27	3.73
Augmented factor input	0.48	2.02	1.24	1.00	2.42	0.79	1.57	5.44	2.95
Foreign trade effect	0.03	0.28	0.18	-0.13	0.36	0.20	0.05	0.38	0.18
Structural effect	0.04	0.46	-0.10	0.20	0.36	0.13	0.40	1.22	0.15
Technology diffusion	0.00	0.25	0.21	0.00	0.34	0.15	0.00	0.50	0.21
Scale	0.03	0.15	0.06	0.04	0.18	0.05	0.07	0.28	0.11
Energy effect	0.00	0.00	-0.05	0.00	0.00	-0.03	0.00	0.00	-0.12
Natural resource windfall	0.00	0.00	0.00	0.00	0.00	0.00	0.00	0.00	0.00
Total explained	0.58	3.19	1.54	1.11	3.69	1.29	2.10	7.97	3.50
Unexplained residual	0.57	1.79	0.61	0.17	2.14	0.50	0.14	1.20	0.23

	Netherlands			UK			USA		
GDP	2.43	4.74	1.78	1.29	3.03	1.75	2.79	3.65	2.51
Augmented factor input	2.09	2.32	1.30	0.94	1.76	0.93	1.53	2.54	2.55
Foreign trade effect	0.15	0.98	0.33	0.01	0.25	0.20	0.04	0.07	0.07
Structural effect	n.a.	−0.07	−0.32	−0.04	0.10	−0.38	0.29[a]	0.12	−0.11
Technology diffusion	0.00	0.23	0.13	0.00	0.07	0.13	0.00	0.00	0.00
Scale	0.07	0.14	0.05	0.04	0.09	0.05	0.08	0.11	0.08
Energy effect	0.00	0.06	−0.25	0.00	−0.01	−0.06	0.00	0.01	−0.18
Natural resource windfall	0.00	0.19	0.00	0.00	0.00	0.14	0.00	0.00	0.00
Total explained	2.31	3.88	1.24	0.95	2.28	1.01	1.94	2.86	2.41
Unexplained residual	0.12	0.83	0.54	0.35	0.73	0.73	0.83	0.77	0.10

[a] 1909–48.

Source: Row 1 from Appendix A. Row 2 from Table 5.10. Rows 3–8 derived as explained in text. Row 9 is derived by multiplying the effects shown in rows 2, 3, 4, 5, 6, 7, and 8. Row 10 is derived by dividing the row 1 effect by the row 9 effect.

TABLE 5.20. *Percentage of Growth which is Explained, 1913–1987*

	Growth			
	1913–50	1950–73	1973–87	1913–87
France	51	63	71	62
Germany	87	62	72	74
Japan	94	86	94	91
Netherlands	95	82	70	82
UK	74	75	58	69
USA	70	87	96	84
Arithmetic average	79	76	77	77

Source: Derived from Table 5.19; each of the first three columns shows the ratio of row 9 i.e. 'Total explained' to GDP growth (row 1). The fourth column is the average of the first three columns.

countries is 71 per cent. It is particularly weak in the case of the USA. The US case is in fact rather mysterious. One distinct possibility is that there has been a slow-down in the rate of technical progress.

EXTENT TO WHICH CONVERGENCE IS EXPLAINED

Table 5.22 presents a supplementary account intended to explain the convergence of the follower countries towards US levels of labour productivity since 1950. As this table (unlike Tables 5.19, 5.20, and 5.21) explains labour productivity (and not GDP), changes in labour input do not figure as an explanatory item, and capital inputs are shown in quantities per man-hour. Each item shows the difference between US performance and that of the individual follower countries.

For the 1950–73 period 73 per cent of convergence is explained for the five countries combined, with the lowest level of explanation for France and Germany and some mild over-explanation for the UK. In 1950–73, the most powerful influence making for convergence was the faster growth of capital stock. All the follower

TABLE 5.21. *Degree to which GDP Growth, Acceleration, and Deceleration are Explained, 1950–1987 (annual average compound growth rates)*

	Acceleration in GDP growth 1950–73 from 1913–50	Explained acceleration from 1950–73	Deceleration in GDP 1973–87	Explained deceleration	% of acceleration explained	% of deceleration explained
France	3.89	2.61	−2.88	−1.65	67	57
Germany	4.64	2.58	−4.12	−2.40	56	58
Japan	7.03	5.87	−5.54	−4.47	70	81
Netherlands	2.31	1.57	−2.96	−2.64	68	89
UK	1.74	1.33	−1.28	−1.27	76	99
USA	0.86	0.92	−1.14	−0.45	107	39
Arithmetic average	3.41	2.50	−2.99	−2.15	74	71

Source: Derived from Table 5.19.

TABLE 5.22. *Identifiable Forces Explaining Labour Productivity Convergence, 1950–1987* (difference in annual average compound growth rate between follower countries and the USA)

	France	Germany	Japan	Netherlands	UK	Average for 5 follower countries
			1950–1973			
Rate of convergence of labour productivity	2.46	3.40	5.02	2.26	0.71	2.77
Total capital input per man-hour	0.74	1.24	1.40	0.92	0.81	1.02
Quality of non-residential capital	0.09	0.04	0.02	0.05	-0.01	0.04
Quality of labour	-0.21	-0.19	0.17	0.04	-0.23	-0.08
Foreign trade effect	0.21	0.29	0.31	0.91	0.18	0.38
Structural effect	0.34	0.24	1.10	-0.19	-0.02	0.29
Technological diffusion	0.25	0.34	0.50	0.23	0.07	0.28
Scale effect	0.04	0.07	0.17	0.03	-0.02	0.06
Energy effect	-0.01	-0.01	-0.01	0.05	-0.02	0.00
Natural resource windfall	0.00	0.00	0.00	0.19	0.00	0.04
Explained convergence	1.45	2.02	3.66	2.23	0.76	2.03
Percentage explained	59	59	73	99	107	73

			1973–1987			
Rate of convergence of labour productivity	2.09	1.53	2.06	1.32	1.28	1.66
Total capital input per man-hour	1.01	0.76	1.44	0.65	0.53	0.88
Quality of non-residential capital	-0.02	-0.12	-0.14	-0.21	-0.09	-0.12
Quality of labour	0.13	-0.25	0.12	0.02	-0.08	-0.01
Foreign trade effect	0.11	0.13	0.11	0.26	0.13	0.15
Structural effect	0.01	0.24	0.26	-0.21	-0.27	-0.01
Technological diffusion	0.21	0.15	0.24	0.13	0.13	0.17
Scale effect	-0.02	-0.03	0.03	-0.03	-0.03	-0.02
Energy effect	0.13	0.15	0.06	-0.07	0.12	0.08
Natural resource windfall	0.00	0.00	0.00	0.00	0.14	0.03
Explained convergence	1.56	1.03	2.09	0.54	0.58	1.16
Percentage explained	75	67	101	41	45	70

Source: Line 1 derived from Table 5.2; line 2 (total capital input per man-hour) derived from Tables 5.3 and 5.10 with non-residential capital weighted 0.23 and residential 0.07; line 3 from Table 5.10 weighted by 0.23; line 4 from Table 5.3 weighted by 0.7; lines 5–10, the supplemental effects, are derived from Table 5.19. Lines 2–11 show the difference between the magnitude for the country specified and the lead country, the USA.

countries enjoyed an investment boom of historic proportions, whereas US capital stock increased much more slowly than it had in 1890–1913 when it was establishing its world leadership position. Other important forces underlying convergence were the foreign trade effect, and accelerated diffusion of technology. The structural effect was most advantageous to Japan, which had the greatest scope for moving labour from low to high productivity sectors. In the Netherlands and the UK, structural effects were less favourable than in the United States, and were a negative factor in the convergence accounts. Other items had much weaker explanatory power. The quality of capital effect was mildly more favourable to the follower countries, but the pace of improvement in labour quality was weaker in three of the follower countries than in the USA. Differences in economies of scale played only a minor role, the average effect of the energy effect was zero, and the natural resource windfall applied only to the Netherlands.

In the 1973–87 period, the average pace of convergence was weaker than in 1950–73, and individual country experience was more closely clustered. This can be seen from the first row in the accounts: in 1950–73 the average pace of convergence for the five followers was 2.77 per cent a year, compared with 1.66 points for the 1973–87 period. In 1950–73 convergence ranged from a rate of 5.02 per cent a year in Japan to 0.71 in the UK; for 1973–87 Japan's convergence rate had fallen to 2.06 per cent a year and the UK's had risen to 1.28 points. The three major elements explaining convergence in 1973–87 were the faster growth of capital stock in the follower countries, the foreign trade effect, and the technological diffusion effect. On average for the five followers the structural impact had virtually no explanatory power, but there were strongly negative effects in the Netherlands and UK, with significant positive effects in Germany and Japan. The quality of capital effect was worse in all the followers than in the USA, as their capital stocks aged at a faster pace, and the average quality of labour effect was less favourable. Scale effects were less favourable in most of the follower countries than in the USA. The adverse effects of the energy problem were less severe in most of the follower countries than in the USA, so that this effect was, on average, the fourth major item explaining convergence.

NOTES ON CHAPTER 5

1. See OECD, *Demographic Trends 1950–1990*, Paris, 1979.
2. For a survey of the literature on growth accounting, see A. Maddison, 'Growth and Slowdown in Advanced Capitalist Economies: Techniques of Quantitative Assessment', *Journal of Economic Literature*, June 1987, pp. 649–98, which includes a more detailed quantitative analysis of the same kind conducted in this chapter, for the same six countries. The approach used here includes a good deal of revision and updating, and uses a *numéraire* in 1985 US prices, rather than the 1980 'international dollars' in the article.
3. See sources cited in A. Maddison, 'Growth and Slowdown', 1987, p. 661, for the rationale of the education adjustment.
4. Schumpeter, in his analysis of business cycles, suggested that innovations come in waves, which are the main cause of irregularity in the pace of advance of best practice productivity. Given the large size and diversity of the US economy and the incremental nature of much innovation, it seems to me that irregularity in the pace of advance is likely to be due mainly to irregularity of the rhythm of investment. Fogel and the modern cliometricians have done a great deal to dedramatize the impact of even such a major innovation as the railway. 'No single innovation was vital for economic growth during the nineteenth century . . . Economic growth was a consequence of knowledge acquired in the course of the scientific revolution in the seventeenth, eighteenth and nineteenth centuries. This knowledge provided the basis for a multiplicity of innovations that was applied to a broad spectrum of economic processes . . . This view makes growth the consequence not of one or a few lucky discoveries but of a broad supply of opportunity created by the body of knowledge accumulated over all preceding centuries': R. W. Fogel, *Railroads and American Economic Growth*, Johns Hopkins, Baltimore, 1964, pp. 234–6.
5. Robert Solow, 'Technical Progress, Capital Formation and Economic Growth', *American Economic Review*, May 1962, pp. 76–86, suggested incorporation of a hypothetical quality improvement in successive vintages of capital on the grounds that physical investment is the prime vehicle by which technical progress is realized. Later he suggested that plausible rates of quality augmentation for Germany or the USA were on the order of 4 or 5 per cent a year, which he derived by assuming that *all* progress was embodied, using a production function where there was no quality improvement for labour, and no disembodied technical progress. E. F. Denison, *Why Growth Rates Differ*, Brookings Institution, Washington, DC, 1967, pp. 144–50, pointed out that such high rates of embodiment were incompatible with what we know about the lives of assets, which would be scrapped earlier than they are if Solow's extreme hypothesis were correct. It is clear that some of the impact of technical progress is disembodied and arises from improvement in work practice and organization and by retrofitting and recombining old assets. However, Solow's basic point is extraordinarily illuminating. Insertion of a modest element of embodied technical progress in the analysis does help explain the nature of the growth process, and clarifies the impact of changes in the age of capital in a way that is not possible outside the vintage context.
6. See W. E. G. Salter, *Productivity and Technical Change*, Cambridge, 1960, for an analysis of the embodiment and age effects.

7. See A. Maddison, 'Growth and Slowdown', 1987, pp. 658–61, for discussion of the weighting problem and the calculation of joint factor input.
8. See G. Dosi, C. Freeman, R. Nelson, G. Silverberg, and L. Soete (eds.), *Technical Change and Economic Theory*, Pinter, London, 1988, for a wide-ranging discussion of these issues.
9. For a discussion of the literature on economies of scale, see A. Maddison, 'Growth and Slowdown', 1987, and P. M. Romer, 'Increasing Returns and Long-Run Growth', *Journal of Political Economy*, 1986, pp. 1002–37. For an empirical study, see Z. Griliches and V. Ringstad, *Economies of Scale and the Form of the Production Function*, North Holland, Amsterdam, 1971.
10. Average OECD import prices are taken from OECD, *Economic Outlook*, various issues. OECD gives prices per barrel, which are multiplied here by 7.4 to arrive at price per ton.

THE ROLE OF POLICY IN ECONOMIC PERFORMANCE SINCE 1950

In the 'golden age' the economic performance of the advanced capitalist countries surpassed all historical records, but it deteriorated significantly after 1973. Average GDP growth fell from 4.8 to 2.6 per cent a year, and the pace of inflation accelerated sharply.

This change in momentum raises three questions of 'ultimate' causality which this chapter tries to answer:

1. To what extent was favourable performance in the 1950s and 1960s due to luck or policy? In so far as it was due to policy, how did this differ from that practised earlier or later?
2. What caused the subsequent deterioration? To what degree was it due to a cluster of accidents and policy mistakes, or to the inevitable termination of previously favourable circumstances?
3. Given the new circumstances, what was the degree of policy success or failure? What was the scope for doing better?

TABLE 6.1. *Indicators of Macroeconomic Performance, 1870–1989*

	1870–1950	1950–73	1973–89
Annual average growth of GDP	2.3	4.9	2.6
Average annual rise in consumer prices	0.1[a]	4.1	7.3

[a] Average for peacetime years.

Sources: Appendices A and E. The figures are arithmetic averages for the 16 countries.

CHARACTERISTICS OF THE GOLDEN AGE (1950-1973)

Several special characteristics enhanced economic performance in the 1950s and 1960s. They were:

1. successful reapplication of liberal policies in international transactions;
2. governmental promotion of buoyant domestic demand;
3. policies and circumstances that kept inflation relatively modest in conditions of very high demand;
4. a backlog of growth possibilities, which made the supply response in Europe and Japan very sensitive to high levels of demand.

The first three characteristics are analysed below; the last was examined in Chapter 5. The burden of the following argument is that success in the golden age was due in considerable measure to enlightened policy, but it was helped by temporarily favourable opportunities for fast growth and modest inflation.

Managed Liberalism in International Transactions

Perhaps the least controversial assertion one can make about the golden age is that it involved a remarkable revival of liberalism in international transactions. Trade and payment barriers erected in the 1930s and during the war were removed. The new-style liberalism was buttressed by effective arrangements for articulate and regular consultation between Western countries and for mutual financial assistance.

Between the wars there was more conflict than co-operation. There were quarrels over war debts and reparations, no regular institutional arrangements for mutual credit or consultation except for the meetings of central bankers at the Bank for International Settlements (BIS). There was a shortage of liquidity, fundamental disequilibrium in exchange rates, and pursuit of beggar-your-neighbour commercial policies after the 1929–31 crisis.

The post-war international payments system provided a workable mechanism for the promotion of freer trade. The adherence to virtually fixed exchange rates was a handicap in some respects but there was an adequate supply of liquidity to meet payments difficulties. Reconstruction was greatly eased by Marshall Aid, which prevented quarrels over war debts and reparations and established the habit of organized mutual consultation and mutual financial support. Trade was freed by abolition of quantitative restrictions in OEEC, reduction of tariffs on a regional basis in the European Community and EFTA, and, more globally, in GATT. Currencies became convertible. The international capital market was reborn.

The results were impressive. In the years 1950–73, the export volume of these countries rose nearly sevenfold. They were a major force sustaining demand and productivity growth and keeping prices in check. In 1938, by contrast, the total export volume of these countries was lower than in 1913.

The creation of the EC attracted a greater amount of direct American private capital to Europe in the 1960s than the US government had provided earlier in Marshall Aid. In 1950 US private direct capital in Europe was valued at $1.7 billion, and by 1973 it had risen to $40 billion. This investment helped to strengthen European productivity and competitiveness. It was a major vehicle for technological transfer from the productivity leader, the USA. It contrasted strongly with the situation in the 1930s, when there was an outflow of capital from Europe to the USA.

Between 1950 and 1973 there was net immigration of 9.4 million people into Western Europe, compared with an outflow of 3.7 million from 1914 to 1949. This eased supply constraints. It facilitated output growth, and moderated inflation.

Managed liberalism added greatly to the buoyancy and resilience of the Western economies. These policies of enlightened self-interest and mutual support were due not only to intelligent digestion of the lessons of the 1930s, but to the urgency of Cold War pressures and to the overwhelming power of the USA to enforce its views in the immediate post-war years.

Governmental Promotion of Domestic Demand Buoyancy

A fundamental innovation in post-war policy was the commitment to full use of resources. In 1950–73 the average unemployment rate in these countries was 2.6 per cent of the labour force, compared with 7.5 per cent in 1920–38 (see Table 6.2). In the USA this new goal was enshrined in the Full Employment Act of 1946, though it was not fully implemented until the 1960s. In the UK

TABLE 6.2. *Indicators of Demand and Price Pressure, 1921–1989 (averages for 16 countries)*

Year	Unemployment rate	Annual change in cost of living index	Year	Unemployment rate	Annual change in cost of living index	Year	Unemployment rate	Annual change in cost of living index
1921	6.2	−7.2	1950	3.2	4.5	1974	2.8	13.2
1922	4.9	−9.2	1951	2.9	13.3	1975	4.0	12.1
1923	3.6	1.0	1952	3.4	6.2	1976	4.4	9.8
1924	3.9	3.4	1953	3.3	0.7	1977	4.7	9.2
1925	4.1	4.4	1954	3.3	1.8	1978	5.0	6.9
1926	5.1	−1.5	1955	2.8	1.5	1979	4.8	7.6
1927	4.8	0.2	1956	2.8	3.5	1980	4.9	10.8
1928	4.3	−0.5	1957	2.8	3.5	1981	5.9	10.2
1929	4.5	0.5	1958	3.3	3.3	1982	7.1	8.6
1930	7.0	−3.4	1959	3.0	1.3	1983	7.8	6.4
1931	10.3	−6.1	1960	2.5	2.2	1984	7.3	5.4
1932	13.7	−5.4	1961	2.3	2.4	1985	6.9	4.8

Year		
1933	13.7	−2.3
1934	11.7	0.5
1935	10.8	0.2
1936	9.5	2.3
1937	8.2	6.2
1938	8.3	3.4
1921–38 average	7.5	−0.6
1962	2.1	3.8
1963	2.1	3.6
1964	1.9	4.0
1965	1.8	4.2
1966	1.9	4.0
1967	2.3	3.7
1968	2.5	4.1
1969	2.2	3.9
1970	2.2	5.2
1971	2.5	5.8
1972	2.7	6.7
1973	2.6	8.4
1950–73 average	2.6	4.2
1986	6.6	3.1
1987	6.5	3.3
1988	6.1	3.6
1989	5.7	4.6
1974–89 average	5.7	7.5

Sources: Appendix C, Table C.6 for unemployment; for 1924–38 the figures are an average for 12 countries (excluding France, Italy, Japan, and Switzerland); for 1921–3 Austria is also excluded. Cost of living changes from Appendix E, Tables E.3 and E.4; for 1921–4 Austria and Germany are excluded, and for 1921 Belgium is excluded. Price averages for whole periods differ from those in Table 6.3 because of rounding.

and Scandinavia, the gospel of fiscal activism and the primordial commitment to full employment propounded in the 1930s by the British economist J. M. Keynes (1883–1946) gained wide acceptance in academic, political, and bureaucratic milieux. It involved fiscal activism augmented from time to time by incomes policy. In France, the objective of full resource use was expressed in more *dirigiste* idiom, with less explicit emphasis on full employment, but with an earlier and stronger commitment to growth and supply-side stimuli in its planning process. Italy and Japan had not participated in the Keynesian tradition, but they also aimed at rapid and ambitious rebuilding of their economies through government intervention to influence both supply and demand. Germany gave greater emphasis to price stability, competition, and work incentives than to buoyant domestic demand, but proclaimed the full employment goal in its Stabilization Law of 1967. In any case, it achieved fuller employment than most countries by export-induced growth.

There was a general move away from the pre-war idea that the budget should be balanced irrespective of the state of the economy. Instead, budgetary strategy came to be viewed as an instrument for promoting macroeconomic equilibrium.

Before the war, great importance was attached to price stability in many, but not all, countries. In the golden age it was often cited as a rhetorical target, but in fact the aim was generally to keep the pace of price increase within limits that did not put too great a strain on international competitiveness. When the outcome of demand management policy was unclear, the tendency was to take the upside risk.

There was a major increase in the proportionate size of public expenditure (see Table 3.17), which gave government a propulsive role in the growth of demand. When combined with the commitment to use fiscal policy for macroeconomic objectives, this transformed the nature of the business cycle, which was dominated by swings in government policy rather than by movements in the private sector. The growth of government revenue and expenditure increased the leverage of discretionary policy. It made automatic stabilizers more potent, and strengthened the buying power of low-income consumers receiving social security benefits.

Government policy played a significant role in stabilizing output,[1] but there were enough errors, lags, and uncertainties to prevent fine tuning. The main achievement was not finesse in stabilization, but success in nurturing a buoyancy of demand which had been created during the war and Marshall Plan period and which kept the economies within a zone of high employment. The clear bias in favour of growth and employment, the lowered attention to risks of price increase or payments difficulties, and the absence of crassly perverse deflationary policies were the most important features differentiating post-war from pre-war domestic policy. The pay-off was bigger than could reasonably have been anticipated. The absence of downside risks in terms of output, and the buoyancy that continuous price increases gave to profits, sparked off a secular investment boom; and, given the favourable supply factors in Europe and Japan, growth performance reached unparalleled proportions.

The Role of Policy and Circumstance in Moderating Price Increases

Those who had contemplated the possibility of achieving full employment in the post-war period had predicted serious inflationary problems,[2] and the outcome was indeed different from previous peacetime experience. The average rate of price increase from 1950 to 1973 was 4.1 per cent a year, compared with 0.4 per cent for 1870–1913, and a fall of 0.7 a year in the inter-war period (see Table 6.3). But in most countries, inflation fell within bounds felt to be tolerable. The gentle upward crawl in prices was considered a reasonable trade-off for low levels of unemployment.

In retrospect, one can identify several reasons for the modesty of price increases in conditions of such high demand. Some of these are related to the exceptional supply factors operative in that period, some to policy-institutional features that will not be easily repeatable. They are as follows.

1. Fixed exchange rates imposed a certain price discipline. It was not as strict as that of the pre-1914 gold standard system, because devaluation was not entirely ruled out, and bigger credits were available to countries in payments difficulty; but balance of payments crises and speculative capital outflows led governments to restrictive action and were felt as constraints by business and trade unions in fixing prices and wages.

2. The rate of inflation in the key currency country had strong international repercussions. From 1952 to the late 1960s US price

TABLE 6.3. *Average Rates of Change in Consumer Price Level, 1870–1989* (annual average compound growth rates)

	1870–1913	1913–20	1920–38	1938–50	1950–73	1973–89
Australia	0.3	8.0	−0.7	4.7	4.6	9.7
Austria	0.1[a]	92.7[c]	2.1[e]	18.1	4.6	4.8
Belgium	0.0	20.0[d]	4.4[f]	11.5	2.9	6.2
Canada	0.6	10.6	−2.4	4.0	2.8	7.4
Denmark	−0.2	15.3	−2.0	5.1	4.8	8.4
Finland	0.6	37.2	0.5	22.3	5.6	8.6
France	0.1	20.5	3.6	28.1	5.0	8.6
Germany	0.6	39.3	−0.1[g]	3.8	2.7	3.6
Italy	0.6	24.4	0.3	38.4	3.9	13.0
Japan	2.8[b]	10.5	−0.3	82.4	5.2	5.3
Netherlands	0.1	9.9	−2.9	7.4	4.1	4.5
Norway	0.6	17.4	−3.1	4.3	4.8	8.5
Sweden	0.5	15.5	−2.7	4.1	4.7	8.6
Switzerland	n.a.	12.5	−2.8	4.0	3.0	3.5
UK	−0.2	13.9	2.6	5.3	4.6	10.3
USA	−0.6	10.1	−2.0	4.5	2.7	6.5
Average	0.4	22.4	−0.7	15.5	4.1	7.3

[a] 1874–1913. [b] 1879–1913. [c] 1914–20. [d] 1914–21. [e] 1923–38. [f] 1921–38. [g] 1924–38.

Source: Appendix E.

increases were very modest. It was not until the Johnson Presidency, with its social programmes and the war in Vietnam, that the economy came under strain.

3. The removal of trade barriers stimulated competition and checked price increases. The average rise in export prices for 1950–73 was only 2.1 per cent a year (see Table 6.4) — half the rise in domestic prices.

4. Key commodity prices were remarkably stable from 1950 to 1970 (see Table 6.5). US agricultural policy created huge stocks, which kept international food prices low or falling. World petroleum prices were stable, due to the large reserves of cheap oil in the Middle East. In non-ferrous metals, sales from the US strategic

TABLE 6.4. *Movement in Export Prices (in National Currencies), 1950–1989* (annual average compound rate of change)

	1950–73	1973–82	1982–9
Australia	0.5	9.9	6.1
Austria	1.8	5.0	0.4
Belgium	1.2	9.1	2.0
Canada	2.2	11.7	−0.8
Denmark	2.3	10.2	2.2
Finland	5.6	14.0	2.3
France	3.8	10.8	4.3
Germany	1.8	5.6	1.1
Italy	0.9	20.0	4.6
Japan	0.9	6.8	−3.5
Netherlands	1.2	9.2	−0.4
Norway	2.0	13.9	−1.1
Sweden	3.0	13.0	4.9
Switzerland	0.5	3.2	2.3
UK	3.5	15.4	3.1
USA	2.7	10.1	−0.4
Average	2.1	10.5	1.7

Sources: First two columns from *Yearbook of International Trade Statistics* and *Monthly Bulletin of Statistics*, UN, New York. Last column from OECD, *Economic Outlook*, various issues.

TABLE 6.5. *Key Commodity Prices, 1950–1989* (yearly averages)

	Gold (London) $ per fine ounce	Petroleum (Saudi Arabia) $ per barrel	Wheat (USA) $ per bushel
1950	35.00	1.71	2.23
1970	35.98	1.30	1.48
1973	100.00	2.70	3.81
1974	102.02	9.76	4.90
1980	607.87	28.67	4.70
1982	375.80	33.47	4.36
1989	381.28	17.18	4.61

Source: IMF, *International Financial Statistics*. The 1989 petroleum price is the average crude price, as Saudi Arabia stopped price quotations.

stockpile put a damper on price movements. Finally, the gold price was stable. Central banks fed the free market with gold up to 1968, which held the price near the official $35 an ounce and helped keep down pressure in speculative markets generally.

5. In the 1950s and 1960s, there was an elastic supply of labour due to immigration and outflows from agriculture. This eased bottle-necks and checked wage increases. The countries that made most explicit use of migrant labour as a cyclical stabilization device were Switzerland and Germany. In Switzerland unemployment was zero and in Germany less than 1 per cent of the labour force in the 1960s, and yet wage pressures were modest. In Italy and Japan surplus rural labour performed the same function.

6. The climate of wage bargaining in the 1950s and most of the 1960s was mild. By pre-war standards, there was a low level of social tension. Several reasons contributed to this: the unprecedented increases in real income; the effect of East–West tensions in consolidating Western societies internally; the solidaristic feelings promoted by wide social security provisions and income transfers. The climate can best be recalled by citing some of the social critics of that epoch. In books published between 1956 and 1960, Crosland, Galbraith, Bell, and Myrdal proclaimed the unimportance of distributive issues, the increasing internal harmony

of Western societies, the rise of legitimate meritocratic élites, etc.[3]

7. Institutions and expectations had not properly adjusted to continuous inflation. People were suffering from 'money illusion', which in the long run would be eroded. Friedman suggested in 1968 that such a phenomenon could well last for a couple of decades, after which expectations would become more rational and decidedly more explosive unless unemployment increased.[4] Indeed, it is rather surprising that money illusion was not broken by wartime experience and by the Korean War boom of 1950–1.

THE BREAKDOWN OF THE GOLDEN AGE

The golden age ended for a complexity of interacting reasons which are not easy to disentangle. Four major elements are distinguished here:

1. the messy collapse of the Bretton Woods fixed exchange rate system;
2. the erosion of price constraints, and the emergence of strong inflationary expectations as a prime element in wage and price determination;
3. the shock of a tenfold rise of oil prices;
4. the erosion of exceptional supply side factors (analysed in Chapter 5).

Except for the timing of the oil shocks, these changes were not due to unusually bad luck, nor were they due to policy error. Rather, there was an erosion of the particularly favourable circumstances of the golden age, which was inevitable in the long run.[5]

The Collapse of the Bretton Woods International Monetary System

In the Bretton Woods system, the dollar was the unit in which other countries kept their reserves, and to which they pegged their exchange rates. The USA started the post-war period with a gold reserve bigger than in all the other countries combined. This was the basis of confidence in the dollar, because the USA was willing and able to sell gold for dollars to foreign central banks. Between 1949, when most countries devalued against the dollar, and August 1971, when President Nixon ended its convertibility into gold, there were very few changes in exchange rates in these

TABLE 6.6. *Total International Reserves, 1950–1987* ($ billion, end-year position)

	1950	1970	1973	1987
Australia	1.5	1.7	6.2	12.6
Austria	0.1	1.8	4.3	17.8
Belgium	0.8	2.8	8.1	26.0
Canada	1.8	4.7	7.3	16.3
Denmark	0.1	0.5	1.5	10.9
Finland	0.1	0.5	0.7	7.4
France	0.8	5.0	15.6	70.6
Germany	0.2	13.6	41.5	125.6
Italy	0.6	5.4	12.2	62.6
Japan	0.6	4.8	13.7	92.8
Netherlands	0.5	3.2	10.4	37.4
Norway	0.1	0.8	1.6	15.0
Sweden	0.3	0.8	2.9	11.1
Switzerland	1.6	5.1	14.3	67.5
UK	3.4	2.8	7.9	50.7
USA	24.3	14.5	33.7	164.5
Total	36.9	67.9	182.0	788.8

Source: IMF, *International Financial Statistics*. The figures include SDRs and IMF positions as well as gold and foreign exchange. The IMF shows foreign currency reserves in terms of SDRs and gold. In revaluing in dollar terms, the London end-year price for gold was used here, i.e. $112.25 in 1973 and $486.24 an ounce in 1987. For 1950 and 1970 gold was valued at $35 an ounce.

countries, so that investment and marketing decisions concerning international trade faced little uncertainty on this score.

In the course of time, the increasing competitiveness in trade of the European countries and Japan led to a weakening of the US payments situation, and the international reserve position changed totally. In 1950 Germany, Italy, and Japan together had reserves of only $1.4 billion; by 1970 they had $23.8 billion. In the same period, US reserves fell from $24.3 billion to $14.5 billion.

The long-run vulnerability of the reserve currency country

became increasingly clear. Its capacity to supply gold for dollars was obviously not going to last unless there was a significant change in exchange rates. In fact, Germany revalued twice by a total of 15 per cent, but the willingness to go further was inhibited by the refusal of other strong currency countries, and particularly Japan, to revalue. The possibility of an official revaluation of gold was not seriously considered in the USA.

In the absence of any convincing evidence of reform in the system, there was a whole series of speculative crises against the prevailing exchange parities. These were all the easier to mount because of the dismantling of exchange controls in Europe and the huge growth in the Eurocurrency market. From negligible levels in the 1950s, the net short-term foreign lending of banks in Europe rose to $57 billion by the end of 1970.[6]

The Bretton Woods system eventually collapsed by unilateral action on the part of the USA, which refused to defend its weak payments situation in 1970–1 in the standard deflationary way. It allowed a huge accumulation of dollars by other countries, and as a proxy devaluation imposed a 10 per cent import surcharge in 1971.[7] The rest of the world was thereby compelled to accept the US devaluation against gold in August 1971. The Smithsonian agreement of December 1971 patched up the fixed exchange system with agreed currency realignments, but this broke down again finally in 1973, because the repeated and perceived inadequacy of fixed-rate realignments in a world where international speculation was so easy made it impossible to defend a fixed rate without having a crisis every few weeks.

Given the freedom for international payments transactions and the differences in national growth and price performance, it is clear in retrospect that the Bretton Woods system of pegged rates could not survive.[8]

The Erosion of Price Constraints

The tactics that the USA used to enforce the devaluation of the dollar had inflationary consequences for other countries. The German situation was described by Otmar Emminger, President of the Bundesbank:

As one who participated in (and was partly responsible for) the decision of West Germany to go over to floating in March 1973, I can testify that the main reason for this decision was the effort to shield the German monetary system against further inflationary foreign exchange inflows, after the central bank had to absorb a dollar inflow worth more than DM 20 billions within five weeks, equivalent to more than double the amount of new central bank money required for a whole year.[9]

The system in its breakdown phase was also inflationary in the deficit countries. This was true of the United States itself, and in the UK there was an almost berserk feeling of liberation from the old constraints of stop–go, which led to the adoption of a wildly expansionist policy:

for two years beginning in September 1971 control over the stock of money in the United Kingdom was non-existent. Bank credit to the private sector rose by 50 per cent, most of it financing either consumer outlays or real estate transactions. Simultaneously, on the fiscal side, total borrowing by the public sector moved from a small negative figure to an annual rate of 6 per cent of GDP.[10]

The collapse of the fixed exchange rate discipline and the subsequent easing of demand management constraints had a sizeable role in the unusually large and synchronized boom in world output in 1972–3.[11] Thanks in part to the easy availability of credit to finance imports, the boom was biggest in the Communist countries and parts of the Third World. In 1973 Soviet GDP rose by 6.8 per cent, Chinese by 12 per cent, and Brazilian by 12.5 per cent.[12]

The boom in output put the normal type of cyclical pressure on the price of manufactured goods, which were in such high demand. But a deeper change had occurred. The once-for-all efficiency gains of trade liberalization were now smaller, and the pressures that the fixed exchange rate regime imposed to shave profit margins had greatly eased with its demise. In 1973–82, the average increase in export prices (10.5 per cent) was higher than the domestic price increase (9.6 per cent — see Tables 6.4 and 6.9).

Price pressures in primary commodity markets were even bigger than for manufactures, again for both cyclical and longer-term reasons. The gold price was now free, and its spectacular rise contributed to speculative fever. The world food price situation also changed drastically after twenty years of stability. During the late 1960s US policy had changed and its stocks dropped.[13] Because of *détente*, the USSR was able to buy large amounts of cereals from the USA, and did so in 1972 when it had a poor harvest. As a result, the price of cereals doubled in 1973. After twenty years close to or under $2 a bushel, wheat rose to $3.80 in 1973 and $4.90 in 1974 (see Table 6.5). All agricultural prices rose in sympathy, and the impact on cost-of-living indices around the world was more or less immediate.

The processes of wage bargaining and price fixing were no longer constrained by money illusion, but were geared much more explicitly to anticipate inflation. Thus, when demand weakened in 1974–5, strong pressures for price increase continued, dominated by inflationary expectations which now acted as a fan instead of a dampener. This was quite different from the situation after the Korean War boom of 1950–1, when expectations quickly relapsed to 'normal'.

The Oil Shock

Until 1973, virtually all the economic problems of these countries were responsive to their domestic policy or amenable to their international co-operation. But the OPEC problem was one over which they had little control. In 1973–4 the price of crude oil jumped to four times its average level in 1972. This was a commodity on which the Western countries had become increasingly dependent in a quarter-century of rapid growth during which oil prices had been remarkably stable. In 1973, oil imports of Western countries were more than seventeen times their 1950 level. Oil in 1973 represented half of energy consumption, compared with a quarter in 1950. Part of the price stability was due to the low cost of extracting oil in the Middle East, and the fact that Saudi Arabia and the Persian Gulf countries were in a semi-colonial situation with prices determined mainly by Western oil companies. It is likely that this situation would have changed in the long run in any case, but the big increases of 1973–4 were sparked off by Arab irritation at US policies in support of Israel. For this reason, the oil price rise was backed by a partial embargo on supplies, which greatly increased its economic repercussions — particularly in Japan, which was most dependent on imported oil.

The OPEC cartel's pricing policy damaged Western economies in six ways. The first was the direct loss in real income because of worsened terms of trade. The second arose from structural changes in prices, demand, and output which brought sudden changes in consumer demand and investment intentions and transmitted powerful temporary forces for recession to the private sector. The third was the direct arithmetic contribution of the oil price rise to inflation. The fourth was its powerful reinforcement to inflationary expectations. The fifth was the creation of payments problems. And the sixth was the uncertainty about how OPEC countries would recycle their trade surpluses.

CHANGED POLICY ASPIRATIONS

The fact that governments found themselves facing new problems of double-digit inflation, floating exchange rates, very open capital markets, balance of payments and terms of trade shocks, led to significant changes in the objectives and tactics of official policy.

In international matters, there was a remarkable degree of fidelity to the neo-liberal policies developed in the golden age and the habit of close international co-operation. But in domestic macropolicy top priority was given to the fight against inflation. It was felt that accommodation of inflation beyond a certain point would lead to hyperinflation, and that hyperinflation would threaten the whole socio-political order. Hence the full employment objective disintegrated.

The Keynesian option of protecting employment and checking prices by incomes policy was discredited by the unsuccessful experience of the Nixon and Heath governments in the USA and UK in the early 1970s, and the other Keynesian option of stemming price increases by fiscal incentives was regarded as a gimmick. There was an ideological swing towards the monetarism of Friedman and later to the views of Hayek and neo-Austrians, who regarded unemployment as a useful corrective.[14] In the 1980s, neo-conservative administrations in the UK and USA attempted to reduce the governmental role in their economies by lowering direct taxes, deregulation, and, in the UK case, privatization of government assets.

In spite of this change in attitudes, the anti-inflationary strategy was a cautious and long-drawn-out affair. In the 1970s, Switzerland was the only country to use strongly deflationary

TABLE 6.7. *Summary Indicators of Fiscal and Monetary Posture, 1960–1989*

	Average fiscal outcome as % of GDP			Long-term government bond yield in real terms		
	1960–73	1974–81	1982–9	1960–73	1974–81	1982–9
France	0.5	−0.9	−2.4	1.64	−0.57	5.37
Germany	0.6	−3.1	−1.7	3.72	2.95	4.83
Japan	1.0[a]	−3.5	−0.7	0.46[b]	−0.76	4.25
Netherlands	−0.5	−2.8	−5.1	1.09	2.04	5.54
UK	−0.8	−3.9	−1.7	2.52	−1.30	4.60
USA	−0.0	−0.9	−2.9	2.04	0.24	5.91
Average	0.1	−2.5	−2.4	1.91	0.43	5.08

[a] 1970–3. [b] 1966–73.

Source: OECD, *Economic Outlook* and *National Accounts*, IMF, *International Financial Statistics*, various issues.

medicine to kill off inflationary expectations. In the early 1980s, the UK emulated Swiss policy, without lasting success.

Fiscal outcomes were more supportive of economic activity than most governments would have wanted. Significant budget deficits emerged (see Table 6.7), and the total size of government increased (Table 3.17). In several countries monetary policy was not tight enough to prevent the emergence of negative interest rates. It was not until the chairman of the US central bank, Paul Volcker, changed Federal Reserve policy and gave a strong upward push to world interest rates that monetary policy generally became tight from 1982 onwards (see Table 6.7). World interest rates remained unusually high throughout the 1980s, due in significant part to fiscal laxity in the USA, which covered its budget deficits by large-scale borrowing on domestic and foreign capital markets.

The budgetary outcomes did not always reflect government volition. Although the expansion of government current and capital spending on goods and services was checked, there was a big increase in transfer expenditures such as unemployment benefits and income support. A good deal of the expansion was due to the built-in stability characteristics of social transfers in advanced welfare states, and a significant part of apparently 'discretionary' increases was the result of social momentum to extend the coverage and generosity of benefits which had built up in earlier years. Higher expenditures on debt service were further 'built-in' factors contributing to the size of the deficits. There were also inconsistencies between the stance on macro- and micropolicy. Under pressure from the relevant interest groups, many governments stepped up intervention to mitigate the unemployment and bankruptcy which macropolicy necessarily involved if it was to achieve its purpose. Some governments, and particularly France, the Netherlands, and Sweden, engaged in costly 'job-sharing' schemes to promote exit from the labour force.

In spite of the budget deficits, it is clear that old Keynesian objectives had been abandoned because unemployment rose in all the countries and reached a peak of 7.8 per cent in 1983. Keynesian critics who urged more expansionary policy, such as Blanchard, Dornbusch, and Layard, have pointed out that the budget deficits were not as big as they looked because of the inflation element in debt repayment and the fact that they would have been smaller on a full employment basis.[15] The only countries where policy reflected old Keynesian attitudes were Norway, which had large oil earnings to offset its payments problems, and Australia in the 1980s, where substantial currency depreciation, inflation, and foreign borrowing were accepted as a consequence of bucking the general trend.

ADEQUACY OF THE RESPONSE TO THE NEW CHALLENGES

The main new policy problems which emerged after 1973 were:

1. meeting the OPEC challenge;
2. breaking inflationary expectations;
3. living with makeshift international payments arrangements in very open economies.

Here we assess the degree of success in meeting these problems and the costs in terms of unemployment and output forgone.

Meeting the OPEC Threat

It was inevitable that the Western economies would be pushed into recession by the OPEC shocks. In the short term, they had no choice but to find some accommodation with them. The longer-run strategy concentrated on reduction of the demand for energy, and stimulating the search for substitutes for OPEC oil. The most powerful instrument favouring these outcomes was of course the huge increase in oil prices which put the appropriate market forces into action (though the USA and Canada frustrated these for some time by price and import controls). More efficient use reduced the average ratio of energy consumption to GDP in these countries by a quarter from 1973 to 1989. Substitute sources for OPEC oil were developed in the North Sea, Mexico, Alaska, and the USSR as well as by expanding use of coal and atomic energy. OPEC production fell by half, which greatly reduced its bargaining power. It took a decade to achieve, but the OPEC challenge was broken, and world oil prices fell by half (see Table 6.5).

Given the large size of the OPEC surpluses in 1974–81, our countries were obliged to have a substantial payments deficit but their cumulative deficit was relatively modest and their foreign exchange reserves rose dramatically (Table 6.6). Germany, Japan,

TABLE 6.8. *Current Balance of Payments as a Percentage of GDP in Current Prices, 1961–1989 (average for years cited)*

	1961–73	1974–81	1982–9		1961–73	1974–81	1982–9
Australia	−2.2	−2.7	−4.8	Japan	0.4	0.2	2.7
Austria	−0.4	−1.9	0.1	Netherlands	0.5	0.8	3.0
Belgium	1.6	−1.3	0.8	Norway	−1.9	−5.1	0.0
Canada	−0.8	−1.7	−0.8	Sweden	0.1	−1.9	−1.3
Denmark	−1.9	−3.4	−3.1	Switzerland	−0.1	3.4	4.5
Finland	−1.5	−2.3	−1.9	UK	0.1	−0.2	−0.5
France	−0.2	−0.1	−0.4	USA	0.4	0.1	−2.3
Germany	0.6	0.5	2.8	Arithmetic average	−0.2	−2.0	−0.1
Italy	1.6	−0.7	−0.6				

Source: OECD, *Economic Outlook*, various issues.

the Netherlands, and Switzerland actually maintained a payments surplus at the height of the OPEC problem. The average payments posture of the group returned to its 1960–73 norm after 1982 (see Table 6.8).

Deceleration of Inflation

Success in tackling the wage–price spiral varied because of differences in socio-economic climate and social attitudes, which came from a complex variety of historical factors. In Germany and Austria memories of hyperinflation had a therapeutic effect on wage and price discipline; the Netherlands had a long post-war record of moderation in wage claims, a social compact atmosphere, and a powerful central bank with *rentier* interests at heart; Switzerland has a long history of price stability, and was willing to make greater sacrifices of output to achieve this goal than any other country; Japan has always had greater wage and price flexibility than the other countries.

The four Scandinavian countries, by contrast, did not set great store by price stability. Their wage–price decision process tends to make them drift with the central tendency in their trading partners.[16] Italy had a decidedly non-consensual atmosphere in wage negotiations, and governments which preferred inflation to confrontation.[17]

In spite of these variations, Table 6.9 shows that all countries had appreciable success in moderating the pace of inflation after the 1974–82 period of maximum strain. The pattern of acceleration and deceleration varied sharply between countries. Germany and Switzerland were the only countries to keep inflation at moderate levels in the period when the OPEC shocks were most sharply felt but since then they have been joined by Austria, Japan, and the Netherlands, which have kept inflation below 3 per cent. After 1982 the rate of inflation in these countries as a whole was reduced to an average of 4.5 per cent a year — near the average for the golden age.

Viability of International Payments Arrangements

Since the collapse of the Bretton Woods system, these countries have had to manage their affairs without a generally agreed international monetary system.

TABLE 6.9. *The Course of Inflation, 1973–1989* (annual average compound rate of change in consumer prices)

	1973–82	1982–9		1973–82	1982–9
Australia	11.4	7.6	Japan	8.4	1.1
Austria	6.3	2.8	Netherlands	6.9	1.4
Belgium	8.2	3.7	Norway	9.9	6.8
Canada	9.8	4.5	Sweden	10.3	6.4
Denmark	11.1	5.0	Switzerland	4.4	2.4
Finland	10.8	5.7	UK	14.5	5.1
France	11.4	5.0	USA	8.7	3.6
Germany	5.1	1.7	Arithmetic average	9.6	4.5
Italy	17.0	8.1			

Source: Appendix E.

They have had to conduct their policy in a world with huge international flows of private capital untrammelled by exchange controls.[18] They have had to live with big and unpredictable fluctuations in exchange rates, persistently large deficits in US payments in the 1980s, and equally persistent surpluses in Japan and Germany. In spite of a much higher level of international reserves, these external uncertainties added to the risks perceived in policies of economic expansion.

A clear dichotomy emerged between the USA, Japan, and the UK, which had floating exchange rates, and most members of the European Community, who built up a zone of monetary stability in the European Monetary System (EMS) after 1978 which created fixed but adjustable rates between its members.

The EMS succeeded in making the objective of exchange rate stability credible to market participants and produced a convergence in inflationary experience amongst its members. This involved substantial costs in countries which had to bring down their inflationary expectations closer to those in Germany — which is effectively the key currency country in this system. The case of France is rather clear. After 1983, France followed restrictive policies in spite of high unemployment and slow growth because it felt that the output forgone would be offset by improved long-run prospects for growth, price stability, and competitiveness.

EMS discipline has been gradually tightened by increasing the degree of fixity in exchange rates and narrowing the bands within which they fluctuate. There is already a substantial degree of policy co-ordination and there are plans to move towards fuller or full monetary union.[19]

The USA, the UK, and Japan preferred floating rates because their financial communities are much more heavily engaged in global financial markets than EMS members. The BIS estimates that in 1989 the daily net turnover in the foreign exchange market was $187 billion in the UK, $119 billion in the USA, and $115 billion in Japan, compared with $26 billion for France or $10 billion for Italy.[20] At the end of 1989 British banks had foreign currrency liabilities of $906 billion, Japanese banks $588 billion, and US banks $334 billion, compared with $63 billion for Germany.[21] These commitments made it more difficult for the three floaters to maintain stable exchange rates.

After 1985 there was an attempt to promote exchange stability between the leading EMS members and the three major floaters. This had only very modest success, because the degree of policy co-ordination was poor and the efficacy of the Louvre agreement had less market credibility than that of the EMS.

It is difficult to envisage any return to a standardized monetary system which covers all of these countries, but, after years of hesitation, the UK joined the exchange rate mechanism of the EMS at the end of 1990, and a substantial zone of exchange stability has been created which makes policy formulation less risky than in the decade after 1973.

THE DEGREE OF POLICY SUCCESS SINCE 1973

It is clear that the oil shocks and the breakdown of Bretton Woods were bound to cause unemployment and interrupt growth. On any reasonable accounting, the most sophisticated governments could be expected to lose output in dealing with these shocks, because they involved new risks for policy, and transition problems in devising and learning how to use new instruments, such as floating exchange rates, or new institutions, like the EMS.

Apart from this, it is clear that some of the special supply-side factors which made for supergrowth in the 1950s and 1960s had eroded. Except in the UK, the efforts to find new 'supply-side' stimuli had little success, and in the US case were counter-productive.

In spite of the shocks, the essential elements of the post-war liberal order were kept intact. The spirit of mutual co-operation and degree of mutual consultation have remained unshaken. In spite of massive debt problems in Latin America and Africa, the international capital market did not collapse, nor have there been the internal financial breakdowns that characterized some countries in the 1930s.

There was success in meeting the OPEC threat and in breaking the momentum of inflationary expectations. There has been progress in rebuilding some kind of international monetary order.

During the 1980s economic growth has been considerably better than in the years of high crisis 1973–81. In the eight years 1981–9 growth of GDP per capita averaged 2.4 per cent a year compared with 1.9 per cent in the previous eight years (see Table 6.10).

TABLE 6.10. *Growth of GDP per Capita, 1973–1981 and 1981–1989 (annual average compound growth rate)*

	1973–81	1981–9		1973–81	1981–9
Australia	1.5	2.0	Japan	2.7	3.5
Austria	2.6	2.2	Netherlands	1.3	1.4
Belgium	1.9	2.2	Norway	3.8	3.4
Denmark	1.0	2.2	Sweden	1.3	2.2
Finland	2.2	3.2	Switzerland	0.5	1.5
France	2.0	1.7	UK	0.6	3.1
Germany	2.0	2.1	USA	1.0	2.3
Italy	3.0	2.2	Arithmetic average	1.9	2.4

Source: Appendix A and B.

Whereas average unemployment increased from 2.6 per cent in 1973 to 7.8 in 1983 (see Table 6.2), it responded to increased growth rates (with a lag), and fell to an average of 5.7 per cent in 1989.

The performance of 1981–9 is probably a better guide to future long-run prospects than what happened in the previous eight years. In the light of historical experience before the golden age, a 2.4 per cent per capita growth rate is indeed encouraging. It is nearly double the 1.3 per cent a year which was achieved in the eight decades from 1870 to 1950.

However, the fact that unemployment remains a good deal higher than in the golden age, suggests that these countries could have achieved a better performance than 2.4 per cent real growth per head. Examination of the inter-country range of performance in Table 6.10 also suggests that some of the slower growth countries could have done better, and that policy for growth could have been bolder.

NOTES ON CHAPTER 6

1. There is evidence that governmental budget activity stabilized output growth in six of the seven countries analysed by B. Hansen and W. W. Snyder, *Fiscal Policy in Seven Countries 1955–1965*, OECD, Paris, Mar. 1969, the only exception being the UK.

2. See W. H. Beveridge, *Full Employment in a Free Society*, Allen & Unwin, London, 1944; R. Nurkse, *The Course and Control of Inflation*, League of Nations, 1946; and A. J. Brown, *The Great Inflation 1939–1951*, Oxford, 1955.

3. See C. A. R. Crosland, *The Future of Socialism*, Jonathan Cape, London, 1956, p. 211; J. K. Galbraith, *The Affluent Society*, Hamish Hamilton, London, 1958; G. Myrdal, *Beyond the Welfare State*, Duckworth, London, 1960; and D. Bell, *The End of Ideology*, Free Press, New York, 1960.

4. See M. Friedman, 'The Role of Monetary Policy', *American Economic Review*, Mar. 1968, p. 11. E. S. Phelps made essentially the same point about adaptive processes in inflationary expectations at about the same time as Friedman: see 'Phillips Curves, Expectations of Inflation and Optimal Unemployment Over Time', *Economica*, Aug. 1967.

5. A different view of the 1970s is reflected in P. McCracken *et al.*, *Towards Full Employment and Price Stability*, OECD, Paris, June 1977, p. 103: 'A key conclusion we draw from this assessment of factors underlying recent experience, is that the most important feature was an unusual bunching of unfortunate events unlikely to be repeated on the same scale, the impact of which was compounded by some avoidable errors in economic policy . . . this upheaval is not necessarily a sign of permanent change to an inevitably more unstable and inflationary world.'

6. See Bank for International Settlements, *Forty-Third Annual Report*, June 1973, pp. 167–71.

7. See S. E. Rolfe and J. L. Burtle, *The Great Wheel: The World Monetary System*, McGraw-Hill, New York, 1973, for a lively description of US policy in this period.

8. The official attempts to reform Bretton Woods were concerned mainly with liquidity problems and the change in the nature of reserves. They had minor success in creating special drawing rights (SDRs) which augmented reserves somewhat: see J. Williamson, *The Failure of World Monetary Reform, 1971–74*, Nelson, London, 1977, for an analysis of official attempts to reform the system.

9. O. Emminger, 'The Exchange Rate as an Instrument of Policy', *Lloyds Bank Review*, July 1979, p. 4.

10. P. M. Oppenheimer, 'Why have General Anti-Inflation Policies not Succeeded?' in E. Lundberg (ed.), *Inflation Theory and Anti-Inflation Policy*, Macmillan, London, 1977, pp. 445–59.

11. The 1973 boom did not have the same intensity in all Western countries. The UK was the most extreme case, with a rise in GDP of 7 per cent — almost three times its post-war norm — whereas in Germany 1973 was a year of normal expansion.

12. See A. Maddison, *The World Economy in the Twentieth Century*, OECD, Paris, 1989.

13. See D. E. Hathaway, 'Food Prices and Inflation', *Brookings Papers on*

   Economic Activity, no. 1, 1974.
14. For an analysis of the change in policy objectives and ideology, see A. Maddison, 'Economic Stagnation, since 1973, its Nature and Causes: A Six Country Survey', *De Economist*, No. 131/4, 1983, particularly pp. 598–604.
15. See O. Blanchard, R. Dornbusch, and R. Layard, *Restoring Europe's Prosperity*, MIT, 1986, pp. 74–5, who adjust the fiscal outcomes shown in our Table 6.7 to take account of the inflation component of public debt service and the cyclical situations.
16. For an interpretation of the Scandinavian inflationary mechanism, see L. Calmfors, *Swedish Inflation and International Price Influences*, Institute for International Economic Studies, Stockholm, Mar. 1975. Both employers and unions accept that wage increases should equal the sum of the world price increase plus domestic productivity growth. Wages in the non-traded sector follow those in the sector producing tradeables. This structuralist explanation of the Scandinavian inflation mechanism was developed simultaneously in Norway and Sweden: see O. Aukrust, 'Inflation in the Open Economy: A Norwegian Model', in L. B. Krause and W. S. Salant (eds.), *World-wide Inflation*, Brookings Institution, Washington, 1977; and G. Edgren, K. O. Faxen, and C. E. Ohdner, *Wage Formation and the Economy*, Allen & Unwin, London, 1973.
17. See F. Giavazzi and L. Spaventa, 'Italy: The Real Effects of Inflation and Disinflation', *Economic Policy*, Apr. 1989, pp. 133–71, who argue that the Italian policy mix in accepting inflation as a trade-off for faster growth was more astute than the opposite policies pursued in the UK.
18. In Dec. 1973 the foreign assets of banks in OECD countries in foreign currencies amounted to $248 billion, compared with official reserves of $182 billion. By the end of 1987, these had risen to $3.056 billion compared with official reserves of $789 billion (see BIS, *46th Annual Report*, Basle, 1976, p. 78, for 1973 bank assets, and BIS, *International Banking and Financial Market Developments*, Dec. 1988, p. 4, for 1987 assets. Exchange reserves are shown in our Table 6.6).
19. See F. Giavazzi, S. Micossi, and M. Miller (eds.), *The European Monetary System*, Cambridge, 1988, for an in-depth analysis of EMS objectives and achievements.
20. See BIS, *60th Annual Report*, Basle, 1990, p. 209.
21. See BIS, *International Banking and Financial Market Developments*, Basle, May 1990, p. 4.

APPENDIX A

SOURCES AND METHODS USED TO MEASURE OUTPUT LEVELS AND GDP GROWTH

The output figures refer, wherever possible, to gross domestic product. This broad aggregate covers the output of the whole economy, and excludes income received from or paid for foreign investment. For 1950 onwards, unless otherwise specified, the figures are derived from currently collected official estimates based on almost identical concepts, as published by OECD. Most of them conform closely to the OECD/UN standardized system.[1]

For years before 1950, the estimates have nearly all been made retrospectively and the underlying data are less complete, particularly for years before 1913. Nevertheless, most of the historical estimates are based on substantial statistical research by distinguished scholars, and in some cases emanate from the governmental statistical service responsible for making the more recent official estimates. The long-term measures are obviously not as comparable as those for 1950 onwards, and in some cases may well be substantially revised when further research is done. The weakest figures for the pre-1913 period are those for Belgium, Italy, the Netherlands, and Switzerland, so the results for these countries should be regarded as tentative.

Many studies of long-term growth simply ignore developments in war years, so there are usually gaps in the series from 1913 to the early 1920s and from 1938 to the late 1940s. However, wartime experience is of considerable relevance to subsequent developments as well as being interesting in itself. These wartime gaps have therefore been filled by rough estimates wherever possible.

The figures on GDP are presented as indices, corrected where necessary to eliminate the effect of territorial change. These changes have generally not been large, but it seemed worth while to eliminate them to improve the comparability of the figures. Ignoring the case of Austria after the First World War, which mainly involved the dismantlement of an empire rather than changes in national boundaries, the biggest changes have occurred in Germany. All the adjustments are described in detail in the country notes.

A major problem in long-term comparisons is the correction for price changes. The choice of different periods as a weighting base can affect quantitative development significantly. Generally, the prices of an earlier year will give a higher increase in output than those of a later year, because of the tendency to consume more of those items whose relative price falls. Thus, if one of the countries had used 1870 weights for the whole period 1870–1970 and another country used 1970 weights, the growth rate in the former would have an upward bias relative to the latter. These problems are not negligible but are mitigated in practice because weighting systems are changed every so often and each national series is made up of a number of separate links.[2]

In order to compare levels of output, income per capita, or productivity in different countries, or to add their output to form an aggregate index, it is useful to have a unit which expresses the comparative value of their currencies better than exchange rates. The latter mainly reflect purchasing power over tradeable goods and services, and are subject to a good deal of fluctuation as a result of capital movements. Eurostat and the OECD regularly make estimates of the purchasing power parity (PPP) of OECD currencies and the United Nations Statistical Office makes such estimates for many other countries. These follow the common methodology of the joint International Comparison Project (ICP). Here the latest available ICP bench-mark estimates (for 1985) were used. Eurostat/OECD make several alternative estimates of PPP using various weighting systems. Their preferred measure which they publish is expressed in terms of 'international dollars', derived by a complicated procedure intended to provide an approximation to world prices. I prefer to measure output at US relative prices, first because this is the price structure to which these countries are converging as their productivity and demand patterns approach US levels; and second because this price structure corresponds to an identifiable reality.[3] An estimate at US relative prices is feasible using the unpublished 'Paasche' PPPs of ICP. These are shown in column 2 of Table A.1 and were supplied by courtesy of Eurostat. Figures for GDP in national currencies in column 1 are from OECD, *National Accounts 1960–1988*. Column 3 shows the result of dividing column 1 by column 2, i.e. it converts the GDP figures into a common and comparable measurement unit, i.e. US dollars at 1985 US relative prices. Column 4 shows the average exchange rates for 1985. In all cases the currencies were undervalued relative to the dollar in 1985. For Switzerland, no PPP estimates were available, and the conversion was made at the exchange rate.

Table A.2 shows the level of output in 1985 US dollars for selected years 1820–1989, corrected to offset the impact of boundary changes. It was derived by merging the results of the 1985 bench-mark estimates of Table A.1 at US relative prices with the time series in Tables A.5 to A.8 at constant national prices. Table A.3 shows the output levels within

TABLE A.1. *1985 Levels of GDP at 1985 US Relative Prices*

	GDP at market prices in million units of national currency	Purchasing power parity units of national currency per US $	GDP in $ million at US relative prices	Exchange rate units of national currency per US $
Australia	226,454	1.15794	195,566	1.432
Austria	1,348,420	15.6754	86,021	20.698
Belgium	4,737,409	41.7626	113,437	59.378
Canada	474,339	1.20922	392,269	1.365
Denmark	615,072	9.1656	67,107	10.596
Finland	334,986	5.67869	58,990	6.198
France	4,700,143	6.75214	696,097	8.985
Germany	1,830,490	2.37162	771,831	2.944
Italy	788,368,000[a]	1197.66	658,258	1909.44
Japan	316,303,000	201.452	1,570,116	238.54
Netherlands	418,180	2.43992	171,391	3.321
Norway	500,199	8.2593	60,562	8.597
Sweden	865,788	7.56799	114,401	8.604
Switzerland	227,950	n.a.	92,776	2.457
UK	354,157	0.532476	665,114	0.779
USA	3,967,472	1.00000	3,967,472	1.000

[a] Adjusted downwards by 3 per cent. See country note for Italy in this Appendix.

Sources: GDP in national currencies and exchange rates from OECD, *National Accounts 1960–88*, Paris, 1990. PPPs (Paasche) supplied by Eurostat. Australia adjusted to a calendar year basis.

Table A.2. *Gross Domestic Product in 1985 US Relative Prices* (adjusted to exclude impact of boundary changes) ($ million)

	Australia	Austria	Belgium	Canada	Denmark	Finland
1820	41	3,320	3,505	n.a.	1,133	747
1870	5,059	6,478	10,640	4,969	2,913	1,636
1890	12,190	10,141	16,173	9,083	4,459	2,671
1913	21,807	18,045	25,036	27,607	8,990	5,227
1929	26,779	18,965	31,420	42,929	13,754	7,610
1938	32,165	18,496	31,320	42,791	16,766	10,736
1950	48,499	19,777	36,527	83,981	22,312	13,950
1960	72,094	35,350	49,070	131,437	30,475	22,580
1973	139,522	65,576	91,706	261,908	52,868	42,332
1987	209,675	88,692	117,418	422,416	68,671	62,628
1989	228,298	95,874	127,958	461,342	69,354	69,141

	France	Germany	Italy	Japan	Netherlands
1820	32,871	14,408	18,164	18,228	3,077
1870	60,397	31,512	33,663	21,290	7,463
1890	77,913	50,481	42,558	32,581	11,672
1913	113,741	103,657	76,873	57,564	19,588
1929	152,868	125,529	100,778	102,924	34,748
1938	147,523	175,284	115,914	141,435	35,786
1950	173,569	166,888	132,802	130,728	47,598
1960	271,273	359,172	239,160	304,629	74,707
1973	537,997	626,607	469,348	1,003,744	137,975
1987	725,897	804,482	695,348	1,676,782	176,684
1989	777,081	867,194	745,235	1,859,145	189,142

	Norway	Sweden	Switzerland	UK	USA
1820	830	2,447	n.a.	28,743	10,106
1870	2,065	5,480	4,924	78,936	89,933
1890	2,950	7,890	6,903	118,403	196,433
1913	5,086	13,770	11,923	176,986	473,332
1929	8,066	18,713	18,421	198,047	771,532
1938	10,584	23,547	19,375	234,507	723,725
1950	14,825	37,399	30,774	284,594	1,311,131

TABLE A.2. (Cont.) *Gross Domestic Product in 1985 US Relative Prices* (adjusted to exclude impact of boundary changes) ($ million)

	Norway	Sweden	Switzerland	UK	USA
1960	21,452	51,417	48,316	377,511	1,805,763
1973	37,020	91,887	84,808	565,655	2,988,621
1987	65,223	120,432	97,316	720,687	4,229,695
1989	69,760	125,844	103,539	770,420	4,556,767

TABLE A.3. *Gross Domestic Product NOT Adjusted for Territorial Change*

	1870	1913	1950	1973	1989
Austria	14,770	48,198	19,777	65,576	95,874
Belgium	10,552	24,836	36,527	91,706	127,958
Canada	4,904	27,245	83,981	261,908	461,342
Denmark	2,788	8,604	22,312	52,868	69,354
France	60,397	108,738	173,569	537,997	777,081
Germany	51,017	174,539	166,888	626,607	867,194
Italy	31,785	75,477	132,802	469,348	745,235
Japan	20,263	57,564	129,533	1,003,744	1,859,145
UK	81,934	183,707	284,594	565,655	770,420
USA	89,669	471,945	1,307,288	2,988,621	4,556,767

the geographic boundaries of the years cited for those countries where frontiers changed.

It should be noted that the relative GDP level of different countries varies according to which ICP bench-mark is used. As the degree of sophistication of the ICP exercise has generally increased over time and its country coverage has increased, I use here the results of the latest 1985 exercise. In *Phases of Capitalist Development*, Oxford, 1982, p. 160, I used PPPs at 1970 US relative prices from ICP II for eight of the countries and proxy PPPs for countries not covered by ICP II. Table A.4 shows the impact of this change in bench-marks on the relative standing of countries covered by ICP II. It is clear from the last column that the 1985 bench-mark reduces the standing of Belgium, France, Germany, and the Netherlands as compared with the 1970 PPPs I used in my 1982 study.

TABLE A.4. *Effect of Change from ICP II to ICP V Benchmark on Relative Standing of Individual Countries*

	Former 1970 bench-mark GDP at 1970 US relative prices ($ million)	Subsequent revision in bench-mark GDP (original = 100)	1970–85 GDP growth in national prices (1970 = 100)	1985 GDP at 1970 US relative prices (cols. 1 × 2 × 3) ($ million)	1985 GDP at 1970 US relative prices (US = 100)	1985 GDP at 1985 US relative prices from Table A.1 (US = 100)	ICP V result compared with adjusted ICP II result for 1985 (ICP II = 100)
Belgium	32,258	99.29	142.98	50,055	3.29	2.86	87.0
France	189,480	100.94	149.27	285,495	18.75	17.55	93.6
Germany	234,052	99.12	138.27	320,776	21.07	19.45	92.3
Italy	154,506	103.95	156.76	251,759	16.54	16.59	100.3
Japan	321,804	98.16	190.74	602,509	39.57	39.57	100.0
Netherlands	46,766	105.68	140.02	69,202	4.55	4.32	95.0
UK	183,066	101.58	133.03	247,388	16.25	16.51	101.6
USA	988,600	102.09	150.86	1,522,506	100.00	100.00	100.0

Source: Column 1 from *Phases of Capitalist Development*, 1982, p. 160. Column 2 is difference between GDP figures for 1970 (in US dollars at current exchange rates) in *National Accounts 1960–88*, OECD, Paris, 1990, p. 126, and those in *Phases*, p. 160 Column 3 is derived from country indices in Table A.8 below. For Italy, see country note and note to Table A.1.

The fact that successive ICP bench-marks show appreciable differences in the relative standing of different countries means that they imply different rates of growth between the survey dates than those shown in the national accounts. This has led Robert Summers and Alan Heston to devise an elaborate compromise technique which purges the inconsistency between growth rates implicit in successive bench-mark estimates and those recorded in national accounts. In their paper, 'A New Set of International Comparisons of Real Product and Prices: Estimates for 130 Countries, 1950–1985', *Review of Income and Wealth*, March 1988, they provide such 'consistentized' estimates using the results of ICP III and IV. Such consistentized estimates can never be definitive because they will need modification to incorporate the results of each successive ICP round which appears. The outstanding virtue of the Summers–Heston article is that it draws attention to an important problem which successive ICP publications have never addressed directly. However, I prefer not to use the Summers–Heston approach because they conflate two different possible sources of error. In my view, changes in coverage, methodology, and degree of error in successive ICP rounds are more likely to be the cause of the problem in these countries than errors in national growth measures. In fact, OECD statisticians have themselves pointed to an error in the 1980 PPPs estimated in ICP IV; see D. Blades and D. Roberts, 'A Note on the New OECD Benchmark Parities for 1985', *OECD Economic Studies*, Autumn 1987, p. 191.

SOURCE NOTES FOR TABLES A.5–A.8: GDP INDICES FOR INDIVIDUAL COUNTRIES

Australia: 1828–60 real GDP derived from N. G. Butlin, 'Contours of the Australian Economy 1788–1860', *Australian Economic History Review*, Sept. 1986, pp. 112–13; 1820–8 real GDP derived from the GNE/GDP ratio in Butlin, 1986, and his real GNE per capita figures in N. G. Butlin, 'Our 200 Years', *Queensland Calendar*, 1988. 1860–1 link derived by using the 1860–1 GDP deflator in W. Vamplew (ed.), *Australians: Historical Statistics*, Fairfax, Broadway, 1987, p. 219. 1861–1938/9 GDP from N. G. Butlin, *Australian Domestic Product, Investment and Foreign Borrowing 1861–1938/39*, Cambridge, 1962, pp. 460–1; amended as indicated in N. G. Butlin, *Investment in Australian Economic Development 1816–1900*, Cambridge, 1964, p. 453, and with revised deflator shown in M. W. Butlin, *A Preliminary Annual Database 1900/01 to 1973/74*, Discussion Paper 7701, Reserve Bank of Australia, May 1977, p. 41. 1938/9–1950 from M. W. Butlin, ibid., p. 85. All figures adjusted to a calendar year basis.

Austria: 1830–1913 from A. Kausel, 'Österreichs Volkseinkommen 1830 bis 1913', in *Geschichte und Ergebnisse der zentralen amtlichen Statistik in Österreich 1829–1979, Beiträge zur österreichischen Statistik*, Heft 550, Vienna, 1979. 1913–50 gross national product, from A. Kausel, N. Nemeth, and H. Seidel, 'Österreichs Volkseinkommen, 1913–63', *Monatsberichte des Österreichischen Instituts für Wirtschaftsforschung*, Sonderheft 14, Vienna, Aug. 1965; 1937–45 from F. Butschek, *Die Österreichische Wirtschaft 1938 bis 1945*, Fischer, Stuttgart, 1979, p. 65. The figures are corrected for territorial change, which has been large. (In 1911–13, present-day Austria represented only 37.4 per cent of the total output of the Austrian part of the Austro-Hungarian Empire.) The figures refer to the product generated within the present boundaries of Austria.

Belgium: 1820–46 movement in agricultural output from estimates supplied by Martine Goossens, 1831–46 industrial output estimates supplied by Jean Gadisseur (1820–31 assumed to increase at same pace as in 1831–42), service output 1820–46 assumed to move with population. 1846–1913 GDP derived from movements in agricultural and industrial output from J. Gadisseur, 'Contribution à l'étude de la production agricole en Belgique de 1846 à 1913', *Revue belge d'histoire contemporaine*, vol. iv, 1–2, 1973, and service output which was assumed to move with employment in services (derived for census years from P. Bairoch, *La Population active et sa structure*, Brussels, 1968, pp. 87–8). 1913 weights and 1913–50 GDP estimates derived from C. Carbonnelle, 'Recherches sur l'évolution de la production en Belgique de 1900 à 1957', *Cahiers Économiques de Bruxelles*, no. 3, Apr. 1959, p. 358. Carbonnelle gives GDP figures for only a few bench-mark years but gives a commodity production series for many more years. Interpolations were made for the service sector to arrive at a figure for GDP for all the years for which Carbonnelle shows total commodity production. Figures corrected to exclude the effect of the cession by Germany of Eupen and Malmédy in 1925, which added 0.81 per cent to population and was assumed to have added the same proportion to output.

Canada: 1851–70 GNP from O. J. Firestone, 'Canada's Changing Economy in the Second Half of the 19th Century', NBER, New York, 1957, processed. 1870–1960 from M. C. Urquhart, 'New Estimates of Gross National Product in Canada 1870 to 1926, along with Official Estimates 1926 to 1985: Some Implications for Canadian Development', IARIW, Rocca di Papa, 1987, processed. Professor Urquhart kindly supplied ratios for conversion from GNP to GDP, and I adjusted the figures for 1949 onwards by 0.987 to offset the incorporation of Newfoundland in that year. 1960 onwards from OECD, *National Accounts*.

Denmark: 1820–1950 GDP at factor cost (1929 prices) from S. A. Hansen, *Økonomisk vækst i Danmark*, vol. ii, Institute of Economic

History, Copenhagen, 1974, pp. 229–32 (figures from 1921 onwards adjusted to offset the acquisition of North Schleswig, which added 5.3 per cent to the population, and 4.5 per cent to GDP).

Finland: 1820–60 per capita GDP assumed to increase by 22.5 per cent, as indicated in S. Keikkinen, R. Hjerppe, Y. Kaukiainen, E. Markkanen, and I. Nummela, 'Förändringar i levnadsstandarden i Finland, 1750–1913', in G. Karlsson (ed.), *Levestandarden i Norden 1750–1914*, Reykjavik, 1987, p. 74. 1860–1973 from R. Hjerppe, *The Finnish Economy 1860–1985: Growth and Structural Change*, Bank of Finland, Helsinki, 1989.

France: 1820–1913 agriculture and services and GDP for 1920–38, and 1949–60 from J.-C. Toutain, *Le Produit intérieur brut de la France de 1789 à 1982, Économies et Sociétés*, Presses Universitaires de Grenoble, 1987, pp. 147–57 and 188. For industrial value added 1820–1913, I used the estimates in M. Lévy-Leboyer and F. Bourguignon, *L'Économie française au XIXe siècle*, Economica, Paris, 1985. Lévy-Leboyer shows slower industrial growth than Toutain, as did F. Crouzet, 'Essai de construction d'un indice annuel de la production industrielle française au XIXe siècle', *Les Annales*, 1970. Interpolation between 1913 and 1920 based on figures for industrial and agricultural output shown in J. Dessirier, 'Indices comparés de la production industrielle et production agricole en divers pays de 1870 à 1928', *Bulletin de la statistique générale de la France, Études speciales*, Oct.–Dec. 1928; service output was assumed stable in this period. Interpolation between 1939 and 1946 was based on A. Sauvy's report on national income to the Conseil Économique, *Journal officiel*, 7 Apr. 1954.[4] The figures from 1918 onwards were adjusted downwards by 4.6 per cent to offset the impact of the return of Alsace-Lorraine; figures for 1861–70 were multiplied by 95.92 to offset inclusion of Alsace-Lorraine; and for 1860 and earlier by 97.65 to offset both the impact of the acquisition of Nice and Savoy in 1861 and the Alsace-Lorraine component.

Germany: 1830–50 GDP estimated from Prussian data in R. H. Tilly, 'Capital Formation in Germany in the Nineteenth Century', in P. Mathias and M. M. Postan (eds.), *Cambridge Economic History of Europe*, vol. vii, I, 1978, pp. 395, 420, and 441. Using 1850 weights for agriculture, industry, and services from W. G. Hoffmann, F. Grumbach, and H. Hesse, *Das Wachstum der deutschen Wirtschaft seit der Mitte des 19. Jahrhunderts*, Springer, Berlin, 1965, p. 454, Prussian per capita output in agriculture and industry were multiplied by population in Germany as a whole. Output in services was assumed to move with population. 1850–1925 net domestic product (value added by industry) at factor cost from Hoffmann, Grumbach, and Hesse, ibid., pp. 454–5. This source gives no figures for 1914–24, but starts again in 1925. The pattern of movement in individual years 1914–24 was derived from annual indices of industrial and agricultural output in Dessirier, 'Indices comparés', using Hoffmann's weights for these sectors and adjusting them to fit his sectoral output bench-marks for 1913 and 1925. Service output was interpolated between Hoffmann's 1913 and 1925 figures for this sector. 1925–39 GDP from *Bevölkerung und Wirtschaft 1872–1972*, Statistical Office, Wiesbaden, 1972, p. 260. 1939–44 GNP from E. F. Denison and W. C. Haraldson, 'The Gross National Product of Germany 1936–1944', *Special Paper 1* (mimeographed) in J. K. Galbraith (ed.), *The Effects of Strategic Bombing on the German War Economy*, US Strategic Bombing Survey, 1945. 1946 from *Wirtschaftsproblemen der Besatzungszonen*, DIW, Duncker & Humblot, Berlin, 1948, p. 135; 1945 was assumed to lie midway between 1944 and 1946. 1947–50 from *Statistics of National Product and Expenditure No. 2, 1938 and 1947 to 1955*, OEEC, Paris, 1957, p. 63. The estimates are fully corrected for territorial change, which was extremely complicated. It can be summarized in simplified form as follows (in terms of ratio of old to new territory): 1870, 96.15 per cent; 1918, 108.39 per cent; 1946, 155.35 per cent.[5]

Italy: For 1861–1951, estimates of GDP at factor cost within present frontiers are available in G. Fua (ed.), *Lo sviluppo economico in Italia*, vol. iii, Angeli, Milan, 1969, pp. 410–12, but use of 1938 weights for the whole period understates growth and makes them less comparable with those for other countries, so I reweighted 1861–1913 with 1870 weights, and 1913–38 with 1913 weights. For the industrial sector I substituted the estimates of Stefano Fenoaltea for the years 1861–1913. These two adjustments raised the growth rate of real GDP by around 0.4 per cent a year for 1861–1913. The reweighting for 1913–38 had little effect. Fenoaltea estimates for mining, utilities, and construction are from 'The Extractive Industries in Italy, 1861–1913', *Journal of European Economic History*, Spring 1988, 'The Growth of the Utilities Industries in Italy 1861–1913', *Journal of Economic History*, Sept. 1982, 'Construction in Italy, 1861–1913', *Rivista di Storia Economica*, International Issue, 1987. For manufacturing I used Fenoaltea, 'Public Policy and Italian Industrial Development 1861–1913', Harvard Ph.D. Thesis, 1967 (Tables 24, 25, and 27) excluding utilities, and adding silk from Fenoaltea, 'The Growth of Italy's Silk Industry, 1861–1913', *Rivista di Storia Economica*, Oct. 1988. For 1951–70 I used estimates of R. Golinelli and M. M. Monterastelli, *Un metodo per la ricostruzione di serie storiche compatibili con la nuova contabilità nazionale*, Prometeia, Bologna, 1990. The official ISTAT estimates I used for 1970 onwards have a more complete coverage of the underground economy than is the case in other countries (20.2 per cent of total GDP). In other countries, underground activities which escape the net of official national accounts statisticians are typically about

3 per cent of GDP (see D. Blades, 'The Hidden Economy and the National Accounts', OECD, *Occasional Studies*, June 1982, p. 39). I therefore made a 3 per cent downward adjustment to the bench-mark level of Italian GDP to enhance international comparability. This has no effect on our index of Italian GDP growth; it only affects the level comparisons. See A. Maddison, 'A Revised Estimate of Italian Economic Growth 1861–1989', *Banca Nazionale del Lavoro Quarterly Review*, June 1991 for further details.

Japan: 1885–1950 from K. Ohkawa and M. Shinohara (eds.), *Patterns of Japanese Development: A Quantitative Appraisal*, Yale, 1979, pp. 278–80 for 1885–1940, pp. 259 for 1940–50, adjusted from GNP to GDP basis using coefficients from pp. 268–9. 1945 GDP taken to be two-thirds of the 1944 level (see Y. Kosai, *The Era of High Speed Growth*, University of Tokyo, 1986, p. 34, for partial indicators for 1945). 1946 GDP assumed to lie half-way between 1945 and 1947. 1950–2 from *National Income White Paper* (in Japanese), 1963 edn., p. 178, adjusted to a calendar year basis. 1952 onwards from OECD, *National Accounts*. An upward adjustment of 0.66 per cent was made for 1946 to offset the impact of territorial change (loss of Okinawa). It was assumed that the official figures published by OECD are adjusted to exclude the effect of the reacquisition of Okinawa in 1973 (which added 0.92 per cent to population and to GDP). The figures for 1820 and 1870 are very rough estimates. For 1870–90, it was assumed that per capita product rose at the same rate as in 1890–1913. For 1820–70 it was assumed that GDP per capita rose by 0.1 per cent a year in line with the position of those who think that there was modest growth in the Tokugawa period — see T. C. Smith, 'Pre-Modern Economic Growth: Japan and the West', *Past and Present*, Aug. 1973, pp. 127–60, and S. B. Hanley and K. Yamamura, *Economic and Demographic Change in Preindustrial Japan, 1600–1868*, Princeton, 1977.[6]

Netherlands: 1580–1700 movement inferred from various pieces of evidence presented in Jan de Vries, *The Dutch Rural Economy in the Golden Age*, Yale, 1974, particularly the evidence on explosive urbanization (pp. 88–9) and the transformation of the rural economy (see Table C.13 below for the 1580 estimates). 1700–1900 derived by linking and interpolating J. L. van Zanden, 'De economie van Holland in de periode 1650–1805: groei of achteruitgang? Een overzicht van bronnen, problemen en resultaten', *Bijdrage en Mededelingen Geschiedenis der Nederlanden*, 1987, p. 607, and J. L. van Zanden, 'Economische groei in Nederland in de negentiende eeuw. Enkele nieuwe resultaten', *Economisch en Sociaal-Historisch Jaarboek*, 1987, pp. 58–60. 1900–60 from C. A. van Bochove and T. A. Huitker, 'Main National Accounting Series, 1900–1986', *CBS Occasional Paper*, no. 17, The Hague, 1987. 1960 onwards from OECD, *National Accounts*.

TABLE A.5. *Movement in GDP, 1700–1869 (1913 = 100.0)*

	Australia	Austria	Belgium	Canada	Denmark	Finland	France	Germany
1820	0.19	[18.4]	14.0		12.6	14.3	28.9	[13.9]
1830	0.71	21.1	16.0		14.5		32.4	15.7
1840	1.95	24.0			16.6		39.4	
1850	4.49	27.8	25.4	[9.2]	22.7		45.3	20.3
1860	15.2	32.1	33.6	13.7	25.3	26.1	52.9	24.9
1861	15.2		34.1		25.7	26.3	49.4	24.3
1862	15.0		35.0		26.5	24.9	53.7	25.5
1863	15.5		36.0		28.2	26.9	55.5	27.4
1864	17.2		37.2		27.9	27.5	56.0	28.2
1865	17.1		37.2		28.9	27.3	54.4	28.3
1866	18.2		38.3		28.9	27.6	54.7	28.5
1867	20.4		38.5		28.9	25.4	51.3	28.6
1868	21.4		39.9		29.4	27.9	56.1	30.3
1869	21.6		41.4		31.1	29.9	57.3	30.5

	Italy	Japan	Netherlands	Norway	Sweden	UK	USA
1700			14.69			4.71	
1780						8.51	
1820	[23.63]	[31.67]	15.71	[16.32]	17.77	16.24	2.135
1830					19.5	18.8	
1840					22.2	23.7	5.333
1850			24.5		25.8	28.2	8.213
1860					33.0	36.4	13.386
1861	39.8				31.1	37.4	
1862	41.0				32.8	37.7	
1863	40.2				34.0	38.0	
1864	41.4				36.1	39.0	
1865	42.5			37.6	34.8	40.2	
1866	44.5			38.3	35.1	40.8	
1867	40.8			39.3	34.5	40.4	
1868	42.6			39.1	35.4	41.7	
1869	43.1			40.7	37.2	42.0	17.647

Note: Figures in square brackets are very rough estimates or inferences. See country notes in this section and Table 1.1 above.

TABLE A.6. *Movement in GDP, 1870–1912 (1913 = 100.0)*

	Australia	Austria	Belgium	Canada	Denmark	Finland	France	Germany
1870	23.2	35.9	42.5	18.0	32.4	31.3	53.1	30.4
1871	22.3	38.5	42.6	18.7	32.5	31.5	53.2	30.2
1872	24.7	38.8	45.2	18.6	34.3	32.6	58.9	32.3
1873	27.3	37.9	45.5	20.3	34.1	34.5	55.2	33.7
1874	28.2	39.6	47.0	20.8	35.1	35.3	62.0	36.2
1875	31.3	39.8	46.9	20.4	35.7	36.0	63.9	36.4
1876	31.2	40.7	47.5	19.2	36.4	38.0	59.9	36.2
1877	32.5	42.1	48.1	20.4	35.4	37.1	61.9	36.0
1878	35.6	43.5	49.5	19.8	36.8	36.4	61.1	37.7
1879	36.1	43.2	50.0	21.8	38.0	36.8	57.9	36.8
1880	38.0	43.8	52.5	22.8	38.9	37.0	61.9	36.5
1881	40.8	45.6	53.2	25.9	39.3	36.0	64.7	37.4
1882	38.5	45.9	55.0	26.9	40.7	39.5	67.3	38.0
1883	44.2	47.8	55.8	26.9	42.1	41.0	66.6	40.1
1884	44.4	49.1	56.3	29.2	42.3	41.3	65.6	41.1
1885	47.3	48.8	57.0	27.5	42.6	42.3	64.7	42.1
1886	47.9	50.4	57.7	27.9	44.3	44.4	64.9	42.4
1887	53.0	53.9	59.9	28.9	45.9	45.1	65.5	44.1
1888	53.3	53.8	60.3	30.7	46.2	46.8	66.7	45.9
1889	57.9	53.3	63.2	31.0	46.8	48.4	66.6	47.2

1890	55.9	56.2	64.6	32.9	49.6	51.1	68.5	48.7
1891	60.2	58.2	64.7	34.0	50.6	50.6	70.1	48.6
1892	52.8	59.5	66.3	33.9	51.8	49.1	71.9	50.6
1893	49.9	59.9	67.3	33.5	52.8	51.0	71.9	53.1
1894	51.6	63.4	68.3	35.2	53.9	55.0	74.6	54.4
1895	48.7	65.1	69.9	35.1	56.9	58.0	72.1	57.0
1896	52.4	66.1	71.3	34.2	59.0	61.8	74.9	59.0
1897	49.5	67.5	72.6	37.9	60.4	64.8	73.5	60.7
1898	57.2	71.3	73.8	39.4	61.4	67.6	77.7	63.3
1899	57.2	72.8	75.3	42.9	64.0	66.0	81.4	65.6
1900	60.6	73.4	77.5	45.3	66.2	69.1	82.9	68.4
1901	58.8	73.7	78.2	49.1	69.0	68.3	80.2	66.8
1902	59.4	76.6	79.8	53.4	70.6	66.9	78.4	68.4
1903	64.1	77.3	81.6	55.4	74.8	71.4	80.8	72.2
1904	68.4	78.5	83.7	56.3	76.4	74.1	82.0	75.1
1905	69.2	82.9	86.1	62.0	77.7	75.3	83.4	76.7
1906	73.9	86.1	87.9	68.6	79.9	78.3	84.0	79.0
1907	76.9	91.4	89.2	72.3	82.9	81.0	88.3	82.5
1908	79.5	91.8	90.1	69.1	85.5	81.9	88.4	83.9
1909	86.0	91.5	91.8	76.0	88.8	85.5	90.9	85.6
1910	92.0	92.8	94.2	82.8	91.5	87.4	87.0	88.7
1911	92.7	95.7	96.4	88.9	96.4	89.9	93.3	91.7
1912	94.9	100.5	98.7	95.7	96.4	94.9	100.8	95.7

TABLE A.6. (Cont.) *Movement in GDP, 1870–1912* (1913 = 100.0)

	Italy	Japan	Netherlands	Norway	Sweden	Switzerland	UK	USA
1870	43.8	37.0	38.1	40.6	39.8	41.3	44.6	19.00
1871	44.3			41.2	40.5		47.0	19.88
1872	43.6			43.8	42.4		47.1	19.88
1873	45.3			44.7	46.3		48.2	21.7
1874	45.2			46.2	48.1		49.0	21.6
1875	46.5			47.6	46.0		50.2	22.7
1876	45.5			49.0	49.0		50.7	23.0
1877	45.5			49.2	47.4		51.2	23.7
1878	46.1			47.7	47.1		51.4	24.7
1879	46.7			48.3	46.3		51.2	27.8
1880	48.9		47.6	49.8	48.5		53.6	31.1
1881	45.6			50.2	48.6		55.5	32.2
1882	49.6			50.0	50.6		57.1	34.2
1883	49.4			49.8	51.1		57.5	35.0
1884	49.8			50.8	52.5		57.6	35.7
1885	50.8	45.6		51.4	52.3		57.3	35.9
1886	53.1	49.5		51.7	52.1		58.2	37.0
1887	54.6	51.5		52.3	52.3		60.5	38.7
1888	54.4	49.3		54.6	53.7		63.2	38.5
1889	52.0	51.9		56.5	56.5		66.6	40.9
1890	55.4	56.6	59.6	58.0	57.3	57.9	66.9	41.5

Year								
1891	55.1	53.9		58.5	58.0		66.9	43.3
1892	52.0	57.5		59.8	59.2		65.3	47.5
1893	54.4	57.7		61.4	59.3		65.3	45.2
1894	53.7	64.6		61.6	60.5		69.7	43.9
1895	54.5	65.5		62.2	63.5		71.9	49.2
1896	56.0	61.9		64.1	67.2		74.9	48.2
1897	53.5	63.2		67.3	69.6		75.9	52.8
1898	58.3	75.2		67.5	71.1		79.6	53.9
1899	59.6	69.6		69.4	72.7	69.5	82.9	58.8
1900	63.0	72.6	74.6	70.6	75.3	71.7*	82.3	60.4
1901	67.0	75.2	72.9	72.5	74.5	73.9*	82.3	67.2
1902	65.2	71.3	76.6	74.0	74.4	76.0*	84.4	67.9
1903	68.3	76.3	78.7	73.7	79.9	78.2*	83.5	71.2
1904	68.9	76.9	78.0	73.6	80.7	80.4*	84.0	70.3
1905	72.8	75.6	80.6	74.5	81.6	82.6*	86.5	75.5
1906	75.5	85.5	84.5	77.2	86.9	84.8*	89.4	84.2
1907	84.0	88.2	84.8	80.1	88.8	86.9*	91.1	85.5
1908	86.0	88.8	84.7	82.7	88.6	89.1*	87.4	78.5
1909	92.7	88.7	88.5	84.9	88.0	91.3*	89.4	88.1
1910	89.3	90.1	89.3	87.9	93.3	93.5*	92.2	89.0
1911	95.1	95.0	91.9	90.6	95.6	95.6*	94.9	91.9
1912	95.9	98.4	97.0	94.7	98.3	97.8*	96.3	96.2

*Interpolations; see country notes for rationale.

TABLE A.7. Movement in GDP, 1913–1949 (1913 = 100.0)

	Australia	Austria	Belgium	Canada	Denmark	Finland	France	Germany
1913	100.0	100.0	100.0	100.0	100.0	100.0	100.0	100.0
1914	92.3	83.5*	93.7*	93.5	106.3	95.6	92.9	85.2
1915	86.9	77.4*	92.5*	100.3	98.9	90.8	91.0	80.9
1916	93.0	76.5*	97.9*	110.2	103.1	92.0	95.6	81.7
1917	95.1	74.8*	84.1*	113.6	97.0	77.3	81.0	81.8
1918	95.6	73.3*	67.8*	106.9	93.8	67.0	63.9	82.0
1919	94.7	61.8*	79.9*	99.6	105.9	80.9	75.3	72.3
1920	101.9	66.4	92.5	99.0	110.9	90.5	87.1	78.6
1921	111.9	73.5	94.1	90.8	107.7	93.5	83.5	87.5
1922	112.7	80.1	103.3	104.1	118.6	103.4	98.5	95.2
1923	115.0	79.3	107.0	110.6	131.1	111.0	103.6	79.1
1924	123.8	88.5	110.5	111.7	131.5	113.9	116.6	92.6
1925	127.9	94.5	112.2	123.2	128.5	120.4	117.1	103.0
1926	126.3	96.1	116.0	130.1	136.0	125.0	120.2	105.9
1927	126.0	99.0	120.3	143.0	138.7	134.8	117.7	116.5
1928	124.0	103.6	126.6	155.7	143.4	143.9	125.9	121.6
1929	122.8	105.1	125.5	155.5	153.0	145.6	134.4	121.1

Year								
1930	116.8	102.2	124.3	150.3	162.1	143.9	130.5	119.4
1931	111.5	94.0	122.1	127.1	163.9	140.4	122.7	110.3
1932	115.7	84.3	116.6	118.0	159.6	139.8	114.7	102.0
1933	122.3	81.5	119.1	109.6	164.7	149.1	122.9	108.4
1934	127.0	82.2	118.1	121.2	169.7	166.0	121.7	118.3
1935	132.2	83.8	125.4	131.0	173.5	173.1	118.6	127.2
1936	138.5	86.3	126.3	138.1	177.8	184.8	123.1	138.4
1937	146.4	90.9	128.0	151.1	182.1	195.3	130.2	153.4
1938	147.5	102.5	125.1	155.0	186.5	205.4	129.7	169.1
1939	147.9	116.2	133.6*	164.3	195.4	196.6	139.0	182.7
1940	157.6	113.2	117.7*	186.9	168.0	186.4	114.7	184.0
1941	175.2	121.3	111.5*	213.0	151.4	192.5	90.7	195.7
1942	195.4	115.2	101.9*	250.7	154.8	193.1	81.3	198.4
1943	202.3	118.0	99.5*	262.0	171.9	215.3	77.2	202.3
1944	195.3	121.0	105.4*	271.9	189.9	215.4	65.2	207.5
1945	185.5	50.0	111.7*	263.5	175.6	202.9	70.7	145.3
1946	178.9	58.4	118.3*	260.8	203.0	219.4	107.5	83.0
1947	183.3	64.4	125.4*	272.3	214.4	224.5	116.5	101.9
1948	195.1	82.0	132.9	277.3	221.5	242.3	125.0	120.8
1949	208.0	97.5	138.3	283.3	231.5	257.0	142.0	140.7

* Interpolations; see counting notes for rationale.

TABLE A.7. (Cont.) *Movement in GDP, 1913–1949* (1913 = 100.0)

	Italy	Japan	Netherlands	Norway	Sweden	Switzerland	UK	USA
1913	100.0	100.0	100.0	100.0	100.0	100.0	100.0	100.0
1914	99.9	97.0	97.3	102.2	99.1	100.1*	101.0	92.3
1915	111.8	106.0	100.6	106.6	99.1	100.1*	109.1	94.9
1916	125.4	122.4	103.3	110.0	97.8	100.7*	111.5	108.0
1917	131.3	126.5	96.7	100.0	85.8	89.7*	112.5	105.3
1918	133.3	127.8	90.7	96.3	84.5	89.4*	113.2	114.8
1919	111.0	140.9	112.4	112.6	89.4	95.3*	100.9	115.8
1920	101.3	132.1	115.8	119.7	94.6	101.5*	94.8	114.7
1921	99.8	146.6	122.9	109.8	91.1	99.0*	87.1	112.1
1922	104.9	146.2	129.6	122.6	99.7	108.5*	91.6	118.3
1923	111.3	146.3	132.8	125.3	105.0	114.8*	94.5	133.9
1924	112.4	150.4	142.5	124.7	108.3	119.1	98.4	138.0
1925	119.8	156.6	148.5	132.4	112.3	127.8	103.2	141.2
1926	121.1	158.0	160.4	135.3	118.6	134.2	99.4	150.4
1927	118.4	160.3	167.1	140.5	122.3	141.4	107.4	151.9
1928	126.9	173.4	176.0	145.1	128.1	149.3	108.7	153.6
1929	131.1	178.8	177.4	158.6	135.9	154.5	111.9	163.0
1930	124.6	165.8	177.0	170.3	138.7	153.5	111.1	147.5

1931	123.9	167.2	166.2	157.1	133.7	147.1	105.4	135.2
1932	127.9	181.2	163.9	167.6	130.1	142.1	106.2	117.1
1933	127.1	199.0	163.6	171.6	132.6	149.2	109.3	114.7
1934	127.6	199.4	160.6	177.1	142.7	149.5	116.5	123.7
1935	139.9	204.9	166.6	184.7	151.8	148.9	121.0	133.6
1936	140.1	219.8	177.1	196.0	160.6	149.4	126.5	152.7
1937	149.7	230.3	187.2	203.0	168.2	156.5	130.9	160.2
1938	150.8	245.7	182.7	208.1	171.0	162.5	132.5	152.9
1939	161.8	284.4	195.1	218.0	182.8	162.3	133.8	165.0
1940	162.8	292.7	171.9	198.6	177.4	164.0	147.2	178.2
1941	160.8	296.7	162.8	203.4	180.4	162.9	160.6	209.6
1942	158.8	295.1	148.8	195.5	191.4	158.8	164.6	249.2
1943	143.8	299.3	145.2	191.6	199.9	157.4	168.2	294.6
1944	116.8	286.4	97.4	181.6	206.7	161.2	161.6	318.7
1945	91.5	143.2*	99.7	203.5	212.3	207.5	154.5	312.8
1946	119.8	155.6*	168.3	225.3	235.6	221.7	147.8	252.9
1947	140.8	168.0	194.8	251.1	241.4	248.4	145.6	245.6
1948	148.8	193.0	215.6	271.1	248.9	253.4	150.2	255.0
1949	159.8	205.9	234.6	276.4	258.0	246.5	155.8	255.2

* Interpolations; see country notes for rationale.

TABLE A.8. *Movement in GDP, 1950–1989 (1913 = 100.0)*

	Australia	Austria	Belgium	Canada	Denmark	Finland	France	Germany
1950	222.4	109.6	145.9	304.2	248.2	266.9	152.6	161.0
1951	231.9	117.1	154.2	321.5	246.4	289.6	162.0	177.8
1952	234.0	117.2	153.0	344.9	251.0	299.3	166.3	193.5
1953	241.3	122.3	157.9	361.0	265.5	301.4	171.1	209.5
1954	256.3	134.8	164.4	358.5	274.7	327.8	179.4	225.6
1955	270.3	149.7	172.2	392.0	273.7	344.5	189.7	252.8
1956	279.6	160.0	177.2	423.7	279.2	354.9	199.3	270.9
1957	285.2	169.8	180.5	436.0	291.3	371.6	211.3	286.1
1958	298.9	176.0	180.3	443.8	299.5	373.6	216.6	296.3
1959	317.3	181.0	186.0	461.8	320.0	395.8	222.8	318.2
1960	330.6	195.9	196.0	476.1	339.0	432.0	238.5	346.5
1961	338.2	206.3	205.8	491.2	360.6	465.0	251.6	362.6
1962	351.7	211.3	216.5	526.1	381.0	478.8	268.4	379.7
1963	376.0	219.9	225.9	553.4	383.5	494.5	282.8	390.2
1964	402.6	233.1	241.6	590.5	419.0	520.5	301.2	416.3
1965	420.9	239.8	250.2	629.6	438.1	548.0	315.6	439.1
1966	439.5	253.3	258.1	672.5	450.1	561.0	332.1	452.0
1967	462.1	261.0	268.1	692.1	465.5	573.3	347.6	451.5

Year								
1968	491.3	272.6	279.4	728.6	484.0	586.4	362.4	476.6
1969	526.4	289.7	298.0	768.1	514.6	642.6	387.8	512.1
1970	555.9	310.4	316.9	788.0	525.0	690.6	410.0	538.5
1971	586.0	326.2	328.5	833.4	539.0	705.1	429.6	554.1
1972	614.1	346.5	345.8	880.8	567.5	758.9	448.6	577.4
1973	639.8	363.4	366.3	948.7	588.1	809.9	473.0	604.5
1974	658.5	377.8	381.3	990.3	582.6	834.4	487.7	606.1
1975	672.4	376.4	375.6	1016.0	578.7	844.0	486.4	596.5
1976	689.6	393.6	396.5	1078.5	616.2	846.3	507.0	628.8
1977	702.7	410.8	398.4	1117.1	626.2	847.3	523.3	647.3
1978	724.8	412.9	409.3	1168.2	635.4	865.7	540.9	665.9
1979	752.7	432.4	418.0	1213.7	658.0	928.6	558.4	693.6
1980	774.8	445.4	436.0	1231.7	655.1	978.1	567.5	703.0
1981	794.7	444.7	431.5	1277.1	649.2	993.5	574.2	704.2
1982	796.3	449.6	438.1	1236.0	668.8	1029.1	588.8	699.6
1983	811.5	459.4	439.8	1275.5	685.7	1059.7	592.9	710.1
1984	855.2	465.2	449.2	1356.7	715.8	1092.2	600.7	730.1
1985	896.8	476.7	453.1	1420.9	746.5	1128.6	612.0	744.6
1986	928.3	482.1	460.4	1464.5	769.4	1152.2	626.1	762.1
1987	961.5	491.5	469.0	1530.1	763.9	1198.2	638.2	776.1
1988	998.3	511.9	489.1	1606.9	760.8	1260.3	660.7	804.7
1989	1046.9	531.3	511.1	1671.1	771.5	1322.8	683.2	836.6

TABLE A.8. (Cont.) *Movement in GDP, 1950–1989 (1913 = 100.0)*

	Italy	Japan	Netherlands	Norway	Sweden	Switzerland	UK	USA
1950	172.8	227.1	243.0	291.5	271.6	258.1	160.8	277.0
1951	185.7	255.4	248.1	305.0	282.4	279.0	166.6	305.4
1952	199.6	285.0	253.1	315.9	286.4	281.3	166.2	317.2
1953	214.0	306.0	275.1	330.5	294.4	291.2	173.9	330.1
1954	225.1	323.3	293.8	346.9	306.8	307.6	180.5	325.6
1955	238.2	351.1	315.6	353.6	315.7	328.3	186.5	343.6
1956	249.0	377.5	327.2	372.1	327.7	350.1	189.4	350.4
1957	263.7	405.1	336.4	382.9	342.4	364.0	193.1	356.1
1958	277.8	428.7	335.4	379.4	344.1	356.3	193.4	353.8
1959	295.1	467.8	351.8	398.9	354.6	378.7	201.1	374.5
1960	311.1	529.2	381.4	421.8	373.4	405.2	213.3	381.5
1961	337.3	592.9	382.5	448.3	394.8	438.0	220.3	392.0
1962	363.6	645.8	408.7	460.9	417.1	459.0	222.6	412.8
1963	389.5	700.5	423.5	478.3	438.1	481.5	231.0	430.5
1964	404.7	782.3	458.6	502.3	479.0	506.8	243.5	455.7
1965	413.8	827.8	482.6	528.8	505.4	522.9	249.7	482.7
1966	435.4	915.9	495.8	548.9	520.1	535.7	254.3	507.4
1967	466.4	1017.4	522.0	583.2	534.6	552.1	260.1	518.9
1968	505.4	1148.4	555.5	596.4	561.0	571.9	270.7	540.6

Year								
1969	534.3	1291.7	591.2	623.2	575.0	604.1	276.3	556.1
1970	546.3	1430.0	624.9	635.7	612.8	642.7	282.5	555.6
1971	555.0	1491.4	651.3	664.8	636.2	668.8	288.1	573.6
1972	569.9	1616.4	672.9	699.1	634.9	690.3	298.1	602.6
1973	610.6	1743.7	704.4	727.9	667.3	711.3	319.5	631.4
1974	643.6	1722.5	732.3	765.7	688.7	721.7	314.1	626.9
1975	626.6	1767.3	731.7	797.6	706.2	664.5	311.6	620.5
1976	667.8	1851.9	769.1	851.9	713.7	659.7	320.2	650.7
1977	690.4	1949.8	786.9	882.4	702.3	675.8	327.7	679.6
1978	715.8	2049.3	806.3	922.5	714.6	678.6	339.6	714.4
1979	758.6	2155.6	825.4	969.2	742.1	695.5	349.1	728.5
1980	790.9	2251.4	832.5	1010.0	759.0	727.5	341.3	728.0
1981	798.3	2338.3	826.7	1018.9	759.3	738.0	336.9	744.4
1982	800.9	2404.6	815.0	1022.3	767.7	729.7	342.7	725.3
1983	810.0	2480.4	826.5	1069.6	781.6	734.7	354.9	753.7
1984	834.3	2604.8	852.6	1131.1	812.7	747.5	362.4	807.7
1985	856.3	2727.6	875.0	1190.8	830.8	778.1	375.8	838.2
1986	878.1	2794.7	892.6	1240.6	849.7	800.0	388.8	861.9
1987	904.2	2912.7	902.0	1282.4	874.6	816.2	407.2	893.6
1988	939.7	3079.2	926.7	1296.4	895.0	842.5	424.0	934.9
1989	969.4	3229.7	965.6	1371.6	913.9	868.4	435.3	962.7

Norway: GDP at market prices. 1865–1950 from *National Accounts 1865–1960*, Central Bureau of Statistics, Oslo, 1965, pp. 348–59 (I adjusted gross fixed investment downwards by a third to eliminate repairs and maintenance). 1939–44 movement in national income (excluding shipping and whaling operations carried out from Allied bases, 1940–4) from O. Aukrust and P. J. Bjerve, *Hva krigen kostet Norge*, Dreyers, Oslo, 1945, p. 45. 1945 assumed to be midway between 1944 and 1946.

Sweden: 1820–1973 figures supplied by Olle Krantz: see Olle Krantz, 'New Estimates of Swedish Historical GDP since the Beginning of the Nineteenth Century', *Review of Income and Wealth*, June 1988.

Switzerland: 1890–1950 real product in international units from C. Clark, *Conditions of Economic Progress* (3rd edn.), Macmillan, London, 1957, pp. 188–9. The rough estimate for 1870 was derived by backward extrapolation of the 1890–1913 movement in output per head. There is a graphical indication of the growth of Swiss real product in F. Kneschaurek, 'Problemen der langfristigen Marktprognose', *Aussenwirtschaft*, Dec. 1959, p. 336. This shows faster growth than C. Clark to 1938. On the other hand, U. Zwingli and E. Ducret, 'Das Sozialprodukt als Wertmesser des langfristigen Wirtschaftswachstums', *Schweizerische Zeitschrift für Volkswirtschaft und Statistik*, Mar.–June 1964, show slower growth for 1910–38 than C. Clark.

UK: 1700–1831 from N. F. R. Crafts, 'British Economic Growth 1700–1831: A Review of the Evidence', *Economic History Review*, May 1983. For 1700–1801 his estimates refer to England and Wales, and I adjusted them to a UK basis by assuming Scottish GDP per head to have been three-quarters of that in England and Wales in 1801 and to have moved parallel to that in England and Wales from 1700 to 1801; it was further assumed that Irish output per head increased at half the rate which Crafts found for England and Wales for 1700–1801. 1801–31 growth shown by Crafts (p. 187) refers to Great Britain (i.e. England, Wales, and Scotland), and was adjusted to a UK basis assuming Irish output per head of population in 1831 to have been half of that in Great Britain (hypothesis of P. Deane, 'New Estimates of Gross National Product for the United Kingdom 1830–1914', *Review of Income and Wealth*, June 1968) and to have been stagnant from 1801 to 1831. 1830–55 gross national product at factor cost from P. Deane, ibid., p. 106, linked to 1855–1950 GDP at factor cost (compromise estimate) from C. H. Feinstein, *National Income Expenditure and Output of the United Kingdom 1855–1965*, Cambridge, 1972, pp. T18–20. Figures from 1920 onwards are increased by 3.8 per cent to offset the exclusion of output in the area which became the Irish Republic.

USA: GDP 1820–40 at 1840 prices derived from P. A. David, 'The Growth of Real Product in the United States before 1840: New Evidence,

Controlled Conjectures', *Journal of Economic History*, June 1967. The method assumes that 1820–40 agricultural output moved parallel with total population; it derives the agricultural productivity movement from this and further assumes that agricultural and non-agricultural productivity grew at the same pace. Agricultural productivity in 1840 is taken as 51 per cent of non-agricultural. 1840–69 movement of gross national product in 1860 prices derived from R. E. Gallman, 'Gross National Product in the United States 1834–1909', *Output, Employment and Productivity in the United States after 1800*, NBER, New York, 1966, p. 26. Gallman does not actually give figures for 1840, 1850, 1860, and 1869: these were extrapolated from neighbouring years. 1869–90 from N. S. Balke and R. J. Gordon, 'The Estimation of Prewar Gross National Product: Methodology and New Evidence', *Journal of Political Economy*, 1989, p. 84. 1890–1929 from J. W. Kendrick, *Productivity Trends in the United States*, NBER, Princeton, 1961, pp. 298–9. 1929–60 GDP from *The National Income and Product Accounts of the United States, 1929–82, Statistical Tables*, US Dept. of Commerce, Sept. 1986, pp. 6, 7 for GNP, pp. 44–5 for income from abroad. 1960 onwards from OECD, *National Accounts*. Figures for years before 1960 exclude Alaska and Hawaii, which added 0.294 per cent to 1960 GDP (see *Survey of Current Business*, Dec. 1980, p. 17). The figures were adjusted to exclude the impact of this geographic change.

Notes on Appendix A

1. The present system is described in *A System of National Accounts*, UN, New York, 1968. The previous system is described in *A Standardised System of National Accounts*, OEEC, Paris, 1959. For our purposes, the two systems are virtually identical.
2. See A. Maddison, 'Measuring European Growth: The Core and the Periphery', in E. Aerts and N. Valerio (eds.), *Growth and Stagnation in the Mediterranean World*, Leuven University Press, 1990, pp. 82–118, for a discussion of these problems.
3. See A. Maddison, 'Comparative Productivity Levels in the Developed Countries', *Banca Nazionale del Lavoro Quarterly Review*, Dec. 1967, for a more detailed exposition of the same point.
4. Sauvy's estimates for this period seem reasonable when checked against estimates of wartime agricultural and industrial output. See M. Cepède, *Agriculture et alimentation en France durant la IIe guerre mondiale*, Genin, Paris, 1961, and *Annuaire de statistique industrielle 1938–1947*, Ministère de l'Industrie et du Commerce, Paris, 1948.
5. See A. Maddison, 'Phases of capitalist development', *Banca Nazionale del Lavoro Quarterly Review*, June 1977, pp. 133–4, for full details.
6. Most of the revisionist work on the Tokugawa period is non-quantitative as far as macroeconomic performance is concerned. However, Susan Hanley has suggested that Japanese living standards were as high as in the UK in 1850, and, inferentially, that Tokugawa growth was much faster than I have assumed: see S. B. Hanley, 'A High Standard of Living in Nineteenth Century Japan: Fact or

Fantasy?', *Journal of Economic History*, Mar. 1983, pp. 183–92. My own estimates suggest that Japanese GDP per capita was only a third of that in the UK in 1850. Although the evidence is poor before 1885, I regard Hanley's conclusions as an aberration of judgement. She virtually ignores subsequent and much firmer evidence which suggests that Japanese GDP per capita did not reach UK levels until the end of the 1970s, and she makes no serious attempt at a binary comparison with the UK in 1850. She gives details of the household budgets of a carpenter, a farmer, and a samurai at varying dates in Tokugawa Japan and, in lieu of a detailed comparison, simply asserts that the 1850 staple British diet consisted mainly of white bread, margarine, and tea. As margarine was not invented until the late 1860s, it is difficult to take her seriously.

APPENDIX B

POPULATION

All figures are adjusted to refer to mid-year, and breaks in the series due to territorial change are indicated in the tables. The figures include all nationals present in the country, armed forces stationed abroad, and merchant seamen at sea. Aliens are included only if they are permanently settled. Unless otherwise specified, the figures for 1950 onwards are from OECD, *Labour Force Statistics*.

Australia: Bureau of Census and Statistics, *Demography Bulletin Yearbooks* to 1957.

Austria: 1960 onwards, OECD, *Labour Force Statistics*; earlier years from *Statistisches Handbuch für die Republik Österreich*, 1975, p. 9; *Statistisches Handbuch für den Bundesstaat Österreich*, Vienna, 1936, p. 21; and A. Kausel, *Österreichs Volkseinkommen 1830 bis 1913*, Statistical Office, Vienna, 1979. The figures refer throughout to the present territory of Austria.

Belgium: Interpolated from *Annuaire statistique de la Belgique et du Congo Belge*, 1955.

Canada: M. C. Urquhart and K. A. H. Buckley, *Historical Statistics of Canada*, Cambridge, 1965, p. 14. Years before 1870 from E. Kirsten, E. W. Buchholz, and W. Kollmann, *Raum und Bevölkerung in der Weltgeschichte*, Ploetz, Würzburg, 1956, and information supplied by R. Marvin McInnis.

Denmark: S. A. Hansen, *Økonomisk vækst i Danmark*, vol. ii, Institute of Economic History, Copenhagen, 1974, pp. 201–4.

Finland: O. Turpeinen, *Ikaryhmittainen Kuolleisuus Suomessa vv. 1751–1970*, Helsinki, 1973.

France: 1861–1950, from *Annuaire statistique de la France*, 1966, pp. 66–72; 1700–1860 from L. Henry and Y. Blayo, 'La Population de la France de 1740 à 1860', *Population*, Nov. 1975, pp. 97–9. The latter source gives figures for five-year intervals which we have interpolated. They referred to the 1861 territory and were therefore multiplied throughout by 98.235 to exclude Savoy and Nice.

Germany: 1946 onwards, from *Statistisches Jahrbuch 1975 für die Bundesrepublik Deutschland*, p. 49 (1950 adjusted to mid-year). Figures refer to Federal Republic including the Saar throughout and also West Berlin. 1817–1913 W. G. Hoffmann, F. Grumbach, and H. Hesse, *Das Wachstum der deutschen Wirtschaft seit der Mitte des 19. Jahrhunderts*, Springer, Berlin, 1965, pp. 172–5 (for 1870–93 these figures differ a little from those in *Bevölkerung und Wirtschaft 1872–1972*, Statistisches Bundesamt, Wiesbaden, 1972, pp. 101–2). 1913–46 information from above sources, *Statistisches Handbuch von Deutschland 1928–1944*, Munich, 1949, p. 18, and I. Svennilson, *Growth and Stagnation in the European Economy*, Economic Commission for Europe, Geneva, 1954, p. 236.

Italy: 1500–1820 derived from K. J. Beloch, *Bevölkerungsgeschichte Italiens*, de Gruyter, Berlin, 1961, pp. 351–4. 1861–1959, resident population from *Sommario di statistiche storiche dell Italia, 1861–1975*, ISTAT, Rome, 1976, adjusted to mid-year.

Japan: Estimates for 1700–1860 supplied by Akira Hayami; I have added 2 million to his preferred estimates for 1500 and 1600, the upper bound he suggests in 'Population Trends in Tokugawa Japan: 1600–1868', International Statistics Congress, 151, 1986. 1873–1950, from *Japan Statistical Yearbook 1975*, pp. 9, 10, and 13, adjusted to mid-year. Figures for 1935–46 include armed forces overseas. Data on armed forces and geographic change taken from I. B. Taeuber, *The Population of Japan*, Princeton, 1958, chapter XVI. It was assumed that in 1946 only half of the overseas forces had been repatriated to Japan.

Netherlands: 1900–50, from *Zeventig jaren statistiek in tijdreeksen*, CBS, The Hague, 1970, p. 14, adjusted to a mid-year basis. 1870–1900, inter-polated from census results from *Jaarcijfers voor Nederland 1939*, The Hague, 1940, p. 4. Earlier years from J. A. Faber *et al.*, 'Population Changes and Economic Developments in the Netherlands: A Historical Survey', *A. A. G. Bijdragen*, vol. 12, 1965, p. 110, and C. A. Oomens, *De loop van de bevolking van Nederland in de negentiende eeuw*, *Statistische Onderzoekingen*, M35, CBS, The Hague, 1989, p. 16.

Norway: *Historical Statistics 1968*, CBS, Oslo, pp. 44–6.

Sweden: *Historical Statistics of Sweden*, Central Statistical Office, Stockholm, 1955, pp. 2–3. Figures for 1820 onwards supplied by Olle Krantz.

Switzerland: 1950 onwards (average for each year including part of seasonal worker population), from OECD, *Labour Force Statistics*, and OECD Statistics Division. Earlier years from *Annuaire statistique de la Suisse*, 1952, pp. 42–3, and K. B. Mayer, *The Population of Switzerland*, Columbia, 1952, pp. 19 and 29.

UK: 1871–1949 from C. H. Feinstein, *National Income Expenditure and Output of the United Kingdom 1855–1965*, Cambridge, 1972, pp.

T120–1, home population except 1915–20 and 1939 onwards when armed forces overseas are included. 1700–1871, England (excluding Monmouth), from E. A. Wrigley and R. S. Schofield, *The Population of England 1541–1871*, Arnold, London, 1981, pp. 533–5. Ireland 1700–1821 derived from D. Dickson, C. O' Gráda, and S. Daultrey, 'Hearth Tax, Household Size, and Irish Population Change 1672–1821', *Proceedings of the Royal Irish Academy*, vol. 82, C, No. 6, Dublin, 1982; and 1821–41 from J. Lee, 'On the Accuracy of the Pre-Famine Irish Censuses', in J. M. Goldstrom and I. A. Clarkson, *Irish Population, Economy, and Society*, Oxford, 1981. Other parts of UK from B. R. Mitchell, *Abstract of British Historical Statistics*, Cambridge, 1962, and P. Deane and W. A. Cole, *British Economic Growth 1688–1959*, Cambridge, 1964.

USA: *Historical Statistics of the United States, Colonial Times to 1970*, US Department of Commerce, 1975, pp. 8 and 1168, resident population except for 1917–19, and 1930 onwards when armed forces overseas are included.

TABLE B.1. *Population, 1500–1860* ('000)

	Australia[a]	Austria[b]	Belgium	Canada[a]	Denmark	Finland	France[c]	Germany
1500	0	1,420	1,400	0	600	300	16,400	12,000
1600	0	1,800	1,600	0	650	400	19,000	16,000
1700	0	2,100	2,000	15	700	400	21,120	15,000
1760	0	2,778	2,530	65	820	490	25,246	18,310
1820	33	3,189	3,397	640	1,097	1,169	30,698	24,905
1830	66	3,538	3,775	1,034	1,209	1,364	32,712	28,045
1840	360	3,716	4,073	1,523	1,289	1,441	34,284	31,126
1850	389	3,950	4,414	2,404	1,424	1,628	35,708	33,746
1860	1,122	4,233	4,649	3,188	1,611	1,738	36,642	36,049

	Italy	Japan	Netherlands	Norway	Sweden	Switzerland	UK	USA[a]
1500	10,000	12,000	950	300	550	650	4,400	0
1600	13,100	14,000	1,500	400	760	1,000	6,800	0
1700	13,200	30,000	1,900	500	1,260	1,200	8,400	251
1760	15,800	30,000	1,960	687	1,916	1,480	11,050	1,594
1820	19,000	31,000	2,355	970	2,585	1,829	21,240	9,618
1830	20,560	32,000	2,638	1,124	2,888	(2,101)	23,935	12,901
1840	22,250	31,000	2,884	1,241	3,139	(2,222)	26,758	17,120
1850	24,080	32,000	3,095	1,392	3,483	(2,379)	27,418	23,261
1860	26,060	33,000	3,326	1,596	3,860	(2,500)	28,840	31,513

[a] Excludes indigenous populations, which were probably as follows before European settlement: Australia, 300; Canada, 1,000; USA, 2,500.

[b] Refers to population within 1989 frontiers.

[c] Excludes territories acquired in 1861 (Savoie, Haute-Savoie, and part of Alpes-Maritimes), which represented 660,000 of the 37,390,000 population of 1861.

Sources: As in the country notes above, supplemented by J. de Vries, *European Urbanization 1500–1800,* Methuen, London, 1984, p. 36; M. Gille, 'The Demographic History of Northern European Countries in the Eighteenth Century', *Population Studies,* June 1949; B. T. Urlanis, *Rost Naselenie v Evrope,* Ogiz, Moscow, 1941, pp. 414–15; J. F. Dewhurst and Associates, *Europe's Needs and Resources,* Twentieth Century Fund, New York, 1961, p. 911; H. E. Driver, *Indians of North America,* Chicago, 1975, p. 63; E. Kirsten, E. W. Bucholz, and W. Kollmann, *Raum und Bevölkerung in der Weltgeschichte,* Ploetz, Würzburg, 1956; and P. Bairoch, 'Europe's Gross National Product 1800–1875', *Journal of European Economic History,* Fall 1976.

TABLE B.2. *Mid-Year Population, Annual Data, 1870–1913* ('000)

	Australia	Austria[a]	Belgium	Canada	Denmark	Finland	France	Germany
1870	1,620	4,520	5,056	3,641	1,793	1,754	38,440	39,231
1871	1,675	4,562	5,096	3,705	1,807	1,786	36,190[b]	40,997[c]
1872	1,722	4,604	5,137	3,772	1,821	1,819	36,140	41,230
1873	1,769	4,646	5,178	3,843	1,838	1,847	36,340	41,564
1874	1,822	4,688	5,219	3,910	1,856	1,873	36,490	42,004
1875	1,874	4,730	5,261	3,968	1,874	1,899	36,660	42,518
1876	1,929	4,772	5,303	4,023	1,894	1,928	36,830	43,059
1877	1,995	4,815	5,351	4,078	1,917	1,957	37,000	43,610
1878	2,062	4,857	5,399	4,136	1,940	1,983	37,180	44,129
1879	2,127	4,899	5,448	4,203	1,960	2,014	37,320	44,641
1880	2,197	4,941	5,497	4,273	1,976	2,047	37,450	45,095
1881	2,269	4,985	5,562	4,338	1,995	2,072	37,590	45,428
1882	2,348	5,030	5,628	4,389	2,013	2,098	37,730	45,719
1883	2,447	5,075	5,694	4,444	2,029	2,130	37,860	46,016
1884	2,556	5,121	5,761	4,500	2,051	2,164	38,010	46,396
1885	2,650	5,166	5,829	4,548	2,076	2,195	38,110	46,707
1886	2,741	5,212	5,872	4,592	2,102	2,224	38,230	47,134
1887	2,835	5,257	5,915	4,639	2,124	2,259	38,260	47,630
1888	2,932	5,303	5,959	4,691	2,143	2,296	38,290	48,168
1889	3,022	5,348	6,003	4,742	2,161	2,331	38,370	48,717

1890	3,107	5,394	6,048	4,793	2,179	2,364	38,380	49,241
1891	3,196	5,446	6,115	4,846	2,195	2,394	38,350	49,762
1892	3,274	5,504	6,182	4,895	2,210	2,415	38,360	50,266
1893	3,334	5,563	6,250	4,943	2,226	2,430	38,380	50,757
1894	3,395	5,622	6,319	4,991	2,248	2,511	38,420	51,339
1895	3,460	5,680	6,388	5,038	2,276	2,483	38,460	52,001
1896	3,523	5,739	6,442	5,086	2,306	2,515	38,550	52,753
1897	3,586	5,798	6,496	5,135	2,338	2,549	38,700	53,569
1898	3,642	5,856	6,552	5,190	2,371	2,589	38,820	54,406
1899	3,691	5,915	6,609	5,247	2,403	2,624	38,890	55,248
1900	3,741	5,973	6,666	5,319	2,432	2,646	38,940	56,046
1901	3,795	6,035	6,747	5,396	2,463	2,667	38,980	56,874
1902	3,850	6,099	6,848	5,507	2,491	2,686	39,050	57,767
1903	3,896	6,164	6,941	5,666	2,519	2,706	39,120	58,629
1904	3,946	6,228	7,030	5,842	2,546	2,735	39,190	59,475
1905	4,004	6,292	7,118	6,010	2,574	2,762	39,220	60,314
1906	4,062	6,357	7,200	6,123	2,603	2,788	39,270	61,153
1907	4,127	6,421	7,280	6,429	2,635	2,821	39,270	62,013
1908	4,197	6,485	7,352	6,640	2,668	2,861	39,370	62,863
1909	4,278	6,550	7,419	6,816	2,702	2,899	39,430	63,717
1910	4,375	6,614	7,438	7,006	2,737	2,929	39,540	64,568
1911	4,500	6,669	7,457	7,222	2,770	2,962	39,620	65,539
1912	4,661	6,724	7,530	7,409	2,802	2,998	39,670	66,146
1913	4,821	6,767	7,605	7,653	2,833	3,027	39,770	66,978

TABLE B.2. (Cont.) *Mid-Year Population, Annual Data, 1870–1913* ('000)

	Italy	Japan	Netherlands	Norway	Sweden	Switzerland	UK	USA
1870	26,262[d]	34,437	3,615	1,735	4,164	2,664	31,393	39,905
1871	27,223	34,648	3,650	1,745	4,186	2,680	31,556	40,938
1872	27,386	34,859	3,693	1,755	4,227	2,697	31,874	41,972
1873	27,540	35,070	3,735	1,767	4,274	2,715	32,177	43,006
1874	27,660	35,235	3,799	1,783	4,320	2,733	32,501	44,040
1875	27,787	35,436	3,822	1,803	4,362	2,750	32,839	45,073
1876	27,993	35,713	3,866	1,829	4,407	2,768	33,200	46,107
1877	28,221	36,018	3,910	1,852	4,457	2,786	33,576	47,141
1878	28,405	36,315	3,955	1,877	4,508	2,803	33,932	48,174
1879	28,579	36,557	4,000	1,902	4,555	2,821	34,304	49,208
1880	28,690	36,807	4,043	1,919	4,572	2,839	34,623	50,262
1881	28,831	37,112	4,091	1,923	4,569	2,853	34,935	51,542
1882	29,061	37,414	4,140	1,920	4,576	2,863	35,206	52,821
1883	29,267	37,766	4,189	1,919	4,591	2,874	35,450	54,100
1884	29,528	38,138	4,239	1,929	4,624	2,885	35,724	55,379
1885	29,806	38,427	4,289	1,944	4,664	2,896	36,015	56,658
1886	30,020	38,622	4,340	1,958	4,700	2,907	36,313	57,938
1887	30,213	38,866	4,392	1,970	4,726	2,918	36,598	59,217
1888	30,408	39,251	4,444	1,977	4,742	2,929	36,881	60,496
1889	30,633	39,688	4,497	1,984	4,761	2,940	37,178	61,775
1890	30,866	40,077	4,545	1,997	4,780	2,951	37,485	63,056
1891	31,059	40,380	4,601	2,013	4,794	2,965	37,802	64,361

Year								
1892	31,261	40,684	4,658	2,026	4,805	3,002	38,134	65,666
1893	31,475	41,001	4,716	2,038	4,816	3,040	38,490	66,970
1894	31,686	41,350	4,774	2,057	4,849	3,077	38,859	68,275
1895	31,865	41,775	4,883	2,083	4,896	3,114	39,221	69,580
1896	32,042	42,196	4,893	2,112	4,941	3,151	39,599	70,885
1897	32,259	42,643	4,954	2,142	4,986	3,188	39,987	72,189
1898	32,469	43,145	5,015	2,174	5,036	3,226	40,381	73,494
1899	32,673	43,626	5,077	2,204	5,080	3,263	40,773	74,799
1900	32,861	44,103	5,142	2,230	5,117	3,300	41,155	76,094
1901	33,054	44,662	5,221	2,255	5,156	3,341	41,538	77,585
1902	33,315	45,255	5,305	2,275	5,187	3,384	41,893	79,160
1903	33,558	45,841	5,389	2,288	5,210	3,428	42,246	80,632
1904	33,812	46,378	5,471	2,297	5,241	3,472	42,611	82,165
1905	34,083	46,829	5,551	2,309	5,278	3,516	42,981	83,820
1906	34,340	47,227	5,632	2,319	5,316	3,560	43,361	85,437
1907	34,611	47,691	5,710	2,329	5,357	3,604	43,737	87,000
1908	34,893	48,260	5,786	2,346	5,404	3,647	44,124	88,709
1909	35,184	48,869	5,842	2,367	5,453	3,691	44,520	90,492
1910	35,519	49,518	5,902	2,384	5,449	3,735	44,916	92,407
1911	35,840	50,215	5,984	2,401	5,542	3,776	45,268	93,868
1912	36,063	50,941	6,068	2,423	5,583	3,819	45,436	95,331
1913	36,167	51,672	6,164	2,447	5,621	3,864	45,649	97,227

[a] Refers to present territory of Austria (frontiers fixed after First World War).
[b] 1871 onwards excludes Alsace and Lorraine, which had 1,570,000 inhabitants in 1870 (according to French accounting).
[c] 1871 onwards includes Alsace and Lorraine, which had 1,574,000 inhabitants in 1870 (according to German accounting).
[d] Rome, population 800,000, is included after 1870; Venice had been added in 1866.

TABLE B.3. *Mid-Year Population, Annual Data, 1913–1949* ('000)

	Australia	Austria[a]	Belgium	Canada	Denmark	Finland[e]	France	Germany[g]
1913	4,821	6,767	7,605	7,653	2,883	3,027	39,770	66,978
1914	4,933	6,806	7,662	7,888	2,866	3,053	39,782	67,790
1915	4,971	6,843	7,697	7,983	2,901	3,083	38,828	67,883
1916	4,955	6,825	7,701	8,006	2,936	3,105	38,255	67,715
1917	4,950	6,785	7,668	8,067	2,972	3,124	37,683	67,368
1918	5,032	6,727	7,599	8,162	3,006	3,125	36,968	66,811
1919	5,193	6,420	7,567	8,331	3,041	3,117	38,700[f]	60,547
1920	5,358	6,455	7,492	8,575	3,079	3,133	39,000	60,894
1921	5,461	6,504	7,444	8,799	3,285[d]	3,170	39,240	61,573
1922	5,574	6,528	7,511	8,927	3,322	3,210	39,420	61,900
1923	5,697	6,543	7,574	9,021	3,356	3,243	39,880	62,307
1924	5,819	6,562	7,646	9,156	3,389	3,272	40,310	62,697
1925	5,943	6,582	7,779[b]	9,307	3,425	3,304	40,610	63,166
1926	6,064	6,603	7,844	9,467	3,452	3,339	40,870	63,630
1927	6,188	6,623	7,904	9,654	3,475	3,368	40,940	64,023
1928	6,304	6,643	7,968	9,851	3,497	3,396	41,050	64,393
1929	6,396	6,664	8,032	10,044	3,518	3,424	41,230	64,739

1930	6,469	6,684	8,076	10,222	3,542	3,449	41,610	65,084
1931	6,527	6,705	8,126	10,387	3,569	3,476	41,860	65,423
1932	6,579	6,725	8,186	10,520	3,603	3,503	41,860	65,716
1933	6,631	6,746	8,231	10,642	3,633	3,526	41,890	66,027
1934	6,682	6,760	8,262	10,750	3,666	3,549	41,950	66,409
1935	6,732	6,761	8,288	10,854	3,695	3,576	41,940	66,871
1936	6,783	6,758	8,315	10,958	3,722	3,601	41,910	67,349
1937	6,841	6,755	8,346	11,054	3,749	3,626	41,930	67,831
1938	6,904	6,753	8,374	11,162	3,777	3,656	41,960	68,558
1939	6,971	6,653	8,392	11,277	3,805	3,686	41,900	69,286
1940	7,042	6,705	8,346	11,392	3,832	3,698	41,000	69,838
1941	7,111	6,745	8,276	11,519	3,863	3,702	39,600	70,244
1942	7,173	6,783	8,247	11,666	3,903	3,708	39,400	70,834
1943	7,236	6,808	8,242	11,808	3,949	3,721	39,000	70,411
1944	7,309	6,834	8,291	11,957	3,998	3,735	38,900	69,865
1945	7,389	6,799	8,339	12,090	4,045	3,758	39,700	67,000
1946	7,474	7,000	8,367	12,314	4,101	3,806	40,290	68,000
1947	7,578	6,971	8,450	12,574	4,146	3,859	40,680	46,992
1948	7,715	6,956	8,557	12,846	4,190	3,912	41,110	48,251
1949	7,919	6,943	8,614	13,469c	4,230	3,963	41,480	49,198

TABLE B.3. (Cont.) *Mid-Year Population, Annual Data, 1913–1949 ('000)*

	Italy[h]	Japan	Netherlands	Norway	Sweden	Switzerland	UK	USA
1913	36,167	51,672	6,164	2,447	5,621	3,864	45,649	97,227
1914	36,443	52,396	6,277	2,472	5,659	3,897	46,049	99,118
1915	36,887	53,124	6,395	2,498	5,696	3,883	46,340	100,549
1916	37,045	53,815	6,516	2,552	5,735	3,883	46,514	101,966
1917	36,892	54,437	6,654	2,551	5,779	3,888	46,614	103,414
1918	36,447	54,886	6,752	2,578	5,807	3,880	46,575	104,550
1919	36,188	55,253	6,805	2,603	5,830	3,869	46,534	105,063
1920	36,336	55,818	6,848	2,635	5,876	3,877	43,718[i]	106,466
1921	38,248	56,490	6,921	2,668	5,929	3,876	44,072	108,541
1922	38,640	57,209	7,032	2,695	5,971	3,874	44,372	110,055
1923	39,051	57,937	7,150	2,713	5,997	3,883	44,596	111,950
1924	39,408	58,686	7,264	2,729	6,021	3,896	44,915	114,113
1925	39,765	59,522	7,366	2,747	6,045	3,910	45,059	115,832
1926	40,105	60,490	7,471	2,763	6,064	3,932	45,232	117,399
1927	40,452	61,430	7,576	2,775	6,081	3,956	45,389	119,038
1928	40,793	62,361	7,679	2,785	6,097	3,988	45,578	120,501
1929	41,076	63,244	7,782	2,795	6,113	4,022	45,672	121,770
1930	41,399	64,203	7,884	2,807	6,131	4,051	45,866	123,188
1931	41,742	65,205	7,999	2,824	6,152	4,080	46,074	124,149
1932	42,040	66,189	8,123	2,842	6,176	4,102	46,335	124,949
1933	42,360	67,182	8,237	2,858	6,201	4,122	46,520	125,690
1934	42,699	68,090	8,341	2,874	6,222	4,140	46,666	126,485
1935	43,032	69,238	8,434	2,889	6,242	4,155	46,868	127,362
1936	43,352	70,171	8,516	2,904	6,259	4,168	47,081	128,181
1937	43,672	71,278	8,599	2,919	6,276	4,180	47,289	128,961
1938	44,026	71,879	8,685	2,936	6,298	4,192	47,494	129,969

1939	44,477	72,364	8,782	2,954	6,326	4,206	47,991	131,028
1940	44,961	72,967	8,879	2,973	6,356	4,226	48,226	132,122
1941	45,352	74,005	8,966	2,990	6,389	4,254	48,216	133,402
1942	45,634	75,029	9,042	3,009	6,432	4,286	48,400	134,860
1943	45,812	76,005	9,103	3,032	6,491	4,323	48,789	136,739
1944	45,926	77,178	9,175	3,060	6,560	4,364	49,016	138,397
1945	46,080	76,224[i]	9,262	3,091	6,636	4,412	49,182	139,928
1946	46,366	76,602[i]	9,424	3,127	6,719	4,467	49,217	141,389
1947	46,040	77,514	9,630	3,165	6,803	4,524	49,519	144,126
1948	46,381	79,527	9,800	3,201	6,884	4,582	50,014	146,631
1949	46,733	81,329	9,956	3,234	6,956	4,640	50,312	149,188

a Refers to present territory of Austria (frontiers fixed after the First World War).

b 1925 onwards includes Eupen and Malmedy, ceded by Germany (population 63,000 in 1925).

c 1949 onwards includes Newfoundland, which added 347,000 in 1949.

d 1921 onwards includes North Slesvig, ceded by Germany (population 164,000 in 1920).

e The 1940 and 1944 treaties which ceded territory to the USSR had no impact on population as all inhabitants were moved to Finland.

f Refers to territory including Alsace-Lorraine from 1919 onwards. 1913–1918 figures including Alsace-Lorraine were 41,690,000, 41,700,000, 40,700,000, 40,100,000, 39,500,000, 38,750,000.

g 1913–18 refers to German Reichsgebiet before post-war losses. Figures for 1919–46 refer to 1937 territory of the Reich, i.e. the Saar is included throughout (though it did not actually return to Germany until 1935). 1947 onwards refers to the territory of the Federal Republic including the Saar throughout (though it was not actually returned until 1959) and also includes West Berlin. 1913–18 figures for the 1937 territory of the Reich would be 60,351,000, 61,081,000, 61,166,000, 61,015,000, 60,701,000, 60,200,000 respectively. There were 1,166,000 people living in the Saar in 1934. In 1938 there were 42,576,000 living in the territory which became the post-war Bundesrepublik (including Saar and West Berlin), and about 46,000,000 in 1946.

h Population of the territory gained in 1921 was 1,474,000. The population of territory lost in 1947 was 630,000.

i 1946 onwards refers to post-war territory excluding Okinawa, Karafuto, and Kuriles. Most of the population of Okinawa remained there whereas that of Karafuto and the Kuriles was mainly repatriated. The loss of population due to boundary changes was therefore about 600,000.

j 1920 onwards excludes Southern Ireland (population 3,103,000 in 1920).

TABLE B.4. *Mid-Year Population, Annual Data, 1950–1989* ('000)

	Australia	Austria	Belgium	Canada	Denmark	Finland	France	Germany[a]
1950	8,177	6,935	8,640	13,737	4,271	4,009	41,836	49,983
1951	8,418	6,936	8,679	14,047	4,304	4,047	42,156	50,528
1952	8,634	6,928	8,731	14,491	4,334	4,091	42,460	50,859
1953	8,821	6,933	8,778	14,882	4,369	4,139	42,752	51,350
1954	8,996	6,940	8,820	15,321	4,406	4,187	43,057	51,880
1955	9,201	6,947	8,869	15,730	4,439	4,235	43,428	52,382
1956	9,421	6,952	8,924	16,123	4,466	4,282	43,843	53,008
1957	9,640	6,966	8,989	16,677	4,488	4,324	44,311	53,656
1958	9,842	6,987	9,053	17,120	4,515	4,360	44,789	54,292
1959	10,056	7,014	9,104	17,522	4,547	4,395	45,240	54,876
1960	10,275	7,048	9,153	17,909	4,581	4,430	45,684	55,433
1961	10,508	7,087	9,184	18,269	4,610	4,461	46,163	56,175
1962	10,700	7,130	9,221	18,615	4,647	4,491	46,998	56,837
1963	10,907	7,175	9,290	18,965	4,684	4,523	47,816	57,389
1964	11,122	7,224	9,378	19,325	4,720	4,549	48,310	57,971
1965	11,341	7,271	9,464	19,678	4,757	4,564	48,758	58,619
1966	11,599	7,322	9,528	20,048	4,797	4,581	49,164	59,148
1967	11,799	7,377	9,581	20,412	4,839	4,606	49,548	59,286
1968	12,009	7,415	9,619	20,729	4,867	4,626	49,915	59,500

1969	12,263	7,441	9,646	21,028	4,893	4,624	50,315	60,067
1970	12,507	7,467	9,651	21,324	4,929	4,606	50,772	60,651
1971	13,067	7,501	9,673	21,595	4,963	4,612	51,251	61,302
1972	13,304	7,544	9,709	21,822	4,992	4,640	51,701	61,672
1973	13,505	7,586	9,739	22,072	5,022	4,666	52,118	61,976
1974	13,723	7,599	9,768	22,395	5,045	4,691	52,460	62,054
1975	13,893	7,579	9,795	22,727	5,060	4,712	52,699	61,829
1976	14,033	7,566	9,811	23,027	5,073	4,726	52,909	61,531
1977	14,192	7,568	9,822	23,295	5,088	4,739	53,145	61,400
1978	14,359	7,562	9,830	23,535	5,104	4,753	53,376	61,327
1979	14,516	7,549	9,837	23,768	5,117	4,765	53,606	61,359
1980	14,695	7,549	9,847	24,070	5,125	4,780	53,880	61,566
1981	14,923	7,565	9,853	24,366	5,122	4,800	54,182	61,682
1982	15,184	7,571	9,856	24,604	5,119	4,827	54,480	61,638
1983	15,393	7,552	9,855	24,803	5,114	4,856	54,729	61,423
1984	15,579	7,553	9,855	24,995	5,112	4,882	54,946	61,175
1985	15,788	7,558	9,858	25,181	5,114	4,902	55,170	61,024
1986	16,018	7,565	9,862	25,374	5,121	4,918	55,393	61,066
1987	16,263	7,575	9,868	25,644	5,127	4,932	55,630	61,077
1988	16,538	7,595	9,844	25,939	5,130	4,946	55,884	61,451
1989	16,833	7,624	9,938	26,248	5,132	4,964	56,160	61,990

[a] Includes Saar and West Berlin throughout.

TABLE B.4. (Cont.) *Mid-Year Population, Annual Data, 1950–1989* ('000)

	Italy	Japan	Netherlands	Norway	Sweden	Switzerland	UK	USA[c]
1950	47,105	82,900	10,114	3,265	7,015	4,694	50,363	152,271
1951	47,418	84,300	10,264	3,296	7,071	4,749	50,574	154,878
1952	47,666	85,600	10,382	3,328	7,125	4,815	50,737	157,553
1953	47,957	86,760	10,494	3,362	7,171	4,877	50,880	160,184
1954	48,299	88,030	10,616	3,395	7,213	4,929	51,066	163,026
1955	48,633	89,060	10,751	3,429	7,262	4,980	51,221	165,931
1956	48,921	89,980	10,888	3,462	7,315	5,045	51,430	168,903
1957	49,182	90,760	11,026	3,494	7,367	5,126	51,657	171,984
1958	49,476	91,580	11,187	3,525	7,415	5,199	51,870	174,882
1959	49,832	92,460	11,348	3,556	7,454	5,259	52,157	177,830
1960	50,198	93,260	11,486	3,585	7,480	5,362	52,373	180,671
1961	50,524	94,100	11,639	3,615	7,520	5,512	52,807	183,691
1962	50,844	94,980	11,806	3,639	7,562	5,666	53,292	186,538
1963	51,199	95,890	11,966	3,667	7,604	5,789	53,625	189,242
1964	51,601	96,900	12,127	3,694	7,662	5,887	53,991	191,889
1965	51,988	97,950	12,292	3,723	7,734	5,943	54,350	194,303
1966	52,332	98,860	12,455	3,753	7,807	5,996	54,643	196,560
1967	52,667	99,720	12,597	3,785	7,869	6,063	54,959	198,712
1968	52,987	101,170	12,730	3,819	7,912	6,132	55,214	200,706
1969	53,317	102,320	12,878	3,851	7,968	6,212	55,461	202,677

1970	53,661	103,720	13,039	3,879	8,043	6,267	55,632	205,052
1971	54,015	104,750	13,194	3,903	8,098	6,343	55,907	207,661
1972	54,400	106,180	13,329	3,933	8,122	6,401	56,079	209,896
1973	54,779	108,660[b]	13,439	3,961	8,137	6,441	56,210	211,909
1974	55,130	110,160	13,545	3,985	8,160	6,460	56,224	213,854
1975	55,441	111,520	13,666	4,007	8,192	6,404	56,215	215,973
1976	55,701	112,770	13,774	4,026	8,222	6,333	56,206	218,035
1977	55,730	113,880	13,856	4,043	8,251	6,316	56,179	220,239
1978	56,127	114,920	13,942	4,060	8,275	6,333	56,167	222,585
1979	56,292	115,880	14,038	4,073	8,294	6,351	56,227	225,055
1980	56,416	116,800	14,150	4,087	8,311	6,385	56,314	227,757
1981	56,503	117,650	14,247	4,100	8,324	6,429	56,379	230,138
1982	56,639	118,450	14,313	4,116	8,327	6,467	56,335	232,520
1983	56,825	119,260	14,367	4,128	8,329	6,482	56,377	234,799
1984	56,983	120,020	14,424	4,141	8,337	6,505	56,488	237,011
1985	57,128	120,750	14,491	4,153	8,350	6,533	56,618	239,279
1986	57,221	121,490	14,572	4,169	8,370	6,573	56,763	241,625
1987	57,331	122,090	14,665	4,187	8,398	6,619	56,930	243,934
1988	57,441	122,610	14,760	4,209	8,436	6,672	57,065	246,329
1989	57,525	123,116	14,849	4,227	8,493	6,723	57,236	248,777

[b] Includes Okinawa prefecture (Ryuku Islands) from 1973 onwards (added a million to population).
[c] Includes Alaska and Hawaii throughout (population 588,000 in 1950).

TABLE B.5. *Net Migration, 1870–1987* ('000)

	Cumulative totals			
	1870–1913	1914–49	1950–73	1974–87
Australia	702	700	2,067	1,086
Belgium	171	219	287	−8
Canada	861	228	2,126	840
France	890	−236	3,630	365
Germany	−2,598	−304[a]	7,070	1,042
Italy	−4,459	−1,771	−2,139	544
Japan	n.a.	197	−72	−129
Netherlands	−121	−29	47	−18
Norway	−589	−129	0	77
Sweden	−895	83	336	167
Switzerland	20	−90	755	6
UK	−6,415	−1,405[b]	−605	15
USA	15,820	6,221	8,257	7,839
Total	3,396	3,684	21,759	11,826

	Average annual flow			
	1870–1913	1914–49	1950–73	1974–87
Australia	16	19	86	78
Belgium	4	6	12	−1
Canada	20	6	88	60
France	20	−7	151	26
Germany	−59	−17[a]	295	74
Italy	−101	−49	−89	39
Japan	n.a.	5	−3	−9
Netherlands	−3	−1	2	−1
Norway	−13	−4	0	6
Sweden	−20	2	14	12
Switzerland	1	−3	31	0
UK	−146	−48[b]	−25	1
USA	360	173	344	560
Total	78	82	906	845

[a] 1922–39 only. [b] excludes 1939–45.

Sources: 1950–87 from OECD, *Labour Force Statistics*, Paris, various issues, and for Australia 1950–8 from *Report of the Committee of Economic Inquiry* (Vernon Report), Canberra, 1965, Table 42. Earlier figures from various official national sources as well as W. Woodruff, *Impact of Western Man*, Macmillan, New York, 1966, p. 108 (Australia); O. J. Firestone, *Canada's Economic Development*, pp. 240–1 (Canada); S. Kuznets, *Economic Growth and Structure*, Harvard, 1965, p. 364 (USA); W. G. Hoffman et al., *Das Wachstum der deutschen Wirtschaft seit der Mitte des 19. Jahrhunderts*, Springer, Berlin, 1965, pp. 173–4 (Germany); B. R. Mitchell, *European Historical Statistics 1750–1970*, Macmillan, London, 1975.

TABLE B.6. *Vital Statistics, 1820–1987*

	Births per 100 population			Years of life expectancy at birth[f]		
	1820	1900	1987	1820	1900	1987
Australia	n.a.	2.82	1.50		57.0	76.1
Austria	4.30	3.50	1.15		40.1	74.1
Belgium	3.23[a]	2.89	1.19		47.1	74.7
Canada	5.69[b]	2.72	1.45			76.7
Denmark	3.15	2.97	1.09		54.6	75.4
Finland	3.66	3.26	1.22		46.7	75.7
France	3.17	2.13	1.38	39.6	47.0	77.2
Germany	3.99	3.56	1.05		46.6	74.8
Italy	n.a.	3.30	0.98		44.5	76.8
Japan	n.a.	3.24	1.11	34.7[g]		78.2
Netherlands	3.50[c]	3.16	1.28	32.2	52.2	76.8
Norway	3.33	2.97	1.29	43.7	56.3	76.8
Sweden	3.30	2.70	1.25	35.2	55.7	76.6
Switzerland	3.29[d]	2.86	1.18	34.5	50.8	77.0
UK	3.03[e]	2.87[h]	1.36	38.7	50.5[h]	75.2
USA	3.52	3.23	1.57		47.3	75.4
Average	3.78	3.01	1.25	36.9	49.7	76.1

[a] 1830. [b] 1820–30 Quebec. [c] 1840. [d] 1831–40. [e] 1839. [f] average for both sexes. [g] 1850. [h] England and Wales.

Sources: Birth-rates 1987 from OECD, *Labour Force Statistics*, Paris; life expectancy 1987 World Bank, *Social Indicators of Development 1989*, Baltimore, 1989. 1820 and 1900 from B. R. Mitchell, *European Historical Statistics 1750–1970*, Macmillan, London, 1975, for most European countries, Canada from M. C. Urquhart and K. A. H. Buckley (eds.), *Historical Statistics of Canada*, Macmillan, Toronto, 1965; Switzerland from K. B. Mayer, *The Population of Switzerland*, Columbia, New York, 1952, p. 75; USA from *Historical Statistics of the United States: Colonial Times to 1970*, US Dept. of Commerce, Washington, DC, 1975. Estimates of life expectancy for years about 1820 are from various national sources and from D. V. Glass and E. Grebenik, 'World Population, 1800–1950', in H. J. Habakkuk and M. Postan, *Cambridge Economic History of Europe*, Cambridge, 1966, p. 73, and for 1900 from B. Mueller, *A Statistical Handbook of the North Atlantic Area*, Twentieth Century Fund, New York, 1965, p. 22.

TABLE B.7. *Mid-Year Population Adjusted to 1989 Boundaries, Benchmark Years, 1820–1989 ('000s)*

	Australia	Austria	Belgium	Canada	Denmark	Finland	France	Germany
1820	33	3,189	3,424	657	1,155	1,169	31,250	15,788
1870	1,620	4,520	5,096	3,736	1,888	1,754	38,440	24,870
1890	3,107	5,394	6,096	4,918	2,294	2,364	40,107	30,014
1913	4,821	6,967	7,666	7,852	2,983	3,027	41,690	40,825
1929	6,396	6,664	8,032	10,305	3,518	3,424	41,230	43,793
1938	6,904	6,755	8,374	11,452	3,777	3,656	41,960	46,376
1950	8,177	6,935	8,640	13,737	4,269	4,009	41,836	49,983
1973	13,505	7,586	9,739	22,072	5,022	4,666	52,118	61,976
1989	16,807	7,618	9,938	26,248	5,132	4,962	56,160	61,990

	Italy	Japan	Netherlands	Norway	Sweden	Switzerland	UK	USA
1820	19,000	31,000	2,355	970	2,585	1,829	19,746	9,656
1870	27,888	34,437	3,615	1,735	4,164	2,664	29,185	40,061
1890	31,702	40,077	4,545	1,997	4,780	2,951	35,000	63,302
1913	37,248	51,672	6,164	2,447	5,621	3,864	42,622	97,606
1929	40,469	63,244	7,782	2,795	6,113	4,022	45,672	122,245
1938	43,419	71,879	8,685	2,936	6,298	4,192	47,494	130,476
1950	47,105	83,662	10,114	3,265	7,015	4,694	50,363	152,271
1973	54,779	108,660	13,439	3,961	8,137	6,441	56,211	211,909
1989	57,525	123,116	14,850	4,228	8,439	6,723	57,202	248,777

APPENDIX C

LABOUR INPUT AND LABOUR PRODUCTIVITY

The basic factors determining growth of the labour supply are:

1. growth in population (as affected by the pattern of births, deaths, and migration);
2. changes in demographic structure;
3. changes in activity rates (i.e. the willingness of potential workers to seek employment);
4. changes in annual working hours per person (which is not usually a matter on which individuals have much choice, as weekly hours, holidays, and vacations are now generally fixed by collective agreements or legislation).

Actual labour input will differ from potential because of variations in the level of demand. The potential labour supply may not be fully used. The variations in demand are most clearly reflected by changes in the level of unemployment, but this is not the only form of labour slack. Weak demand in labour markets may also cause working hours, activity rates, migration, or productivity to fall below trend levels.

This annex indicates the ways in which the above components have influenced the growth of labour input since 1870.

POPULATION GROWTH AND DEMOGRAPHIC STRUCTURE

Population growth experience has varied a good deal in the past century. From 1870 to 1987 the fastest growing country, Australia, increased its population tenfold, whereas France, the slowest growing country, experienced a rise of less than a half.

By contrast with the variety in population growth experience, changes in demographic structure have been strikingly similar over the long term.

1. The child population (below 15 years of age) decreased proportionately in all the countries, and in virtually all cases the decline was very substantial. In 1870 the average child population was more than a third of the total and in 1987 it was less than a fifth.
2. The proportion of old people (65 and over) rose substantially in all the countries, with the sixteen-country average rising from 5 to 14 per cent of the population.
3. The proportion of people of 'working age' (15 to 64 years) has

Table C.1. *A Comparison of Demographic Structures in 1870 and 1987 (% of total population)*

	Population aged 0–14		Population aged 65 and over		Population aged 15–64			
	1870	1987	1870	1987	1870	1913	1950	1987
Australia	42.3	22.6	1.8	10.7	55.9	63.9	65.3	66.7
Austria	33.8	17.6	3.9	14.8	62.2	62.4	66.8	67.6
Belgium	31.9	18.3	6.4	14.3	61.7	64.7	68.1	67.4
Canada	41.6	21.2	3.7	10.9	54.7	62.0	62.6	67.9
Denmark	33.4	17.7	5.8	15.4	60.8	60.2	64.7	66.9
Finland	34.7	18.3	4.7	12.9	60.6	59.2	63.4	67.8
France	27.1	20.6	7.4	13.5	65.5	65.8	65.9	65.9
Germany	34.0	14.9	4.6	15.1	61.4	63.1	67.1	70.0
Italy	32.5	16.3	5.1	14.1	62.4	60.3	65.5	69.6
Japan	33.7	20.4	5.3	10.8	61.0	59.9	59.4	68.8
Netherlands	33.6	18.6	5.5	12.4	60.9	60.1	63.0	69.0
Norway	35.3	19.3	6.2	16.1	58.5	57.9	66.0	64.6
Sweden	34.1	17.9	5.4	17.7	60.5	60.6	66.3	64.4
Switzerland	31.5	17.0	5.5	14.2	63.0	64.0	66.8	68.8
UK	36.1	18.9	5.0	15.5	58.9	64.1	66.9	65.6
USA	39.2	21.5	3.0	12.2	57.8	63.7	65.0	66.3
Average	34.7	18.9	5.0	13.8	60.4	62.0	65.2	67.3

increased everywhere from an average of 60 per cent in 1870 to 67 per cent in 1987.

These changes have not been steady. There was, for instance, a temporary decline in the proportion of working age as a result of the post-war baby boom; but the long-run changes have been remarkably clear and similar (see Table C.1).

ACTIVITY RATES

The hard core of the labour force has always consisted of adult males, whose activity rates have fallen over time, but activity rates for women

TABLE C.2. *Females as a Proportion of the Labour Force, 1910–1987* (% of total labour force)

	1910	1950	1973	1987
Australia	23.4[a]	22.4	33.6	39.7
Austria	35.9	38.5	38.6	40.1
Belgium	30.9	27.9	33.5	40.7
Canada	13.4[b]	21.4	35.3	46.8
Denmark	31.3[b]	33.7	40.5	45.9
Finland	36.5	40.7	45.2	47.1
France	35.6[a]	36.0	36.8	42.5
Germany	30.7[c]	35.1	36.9	39.3
Italy	31.3[b]	25.5	29.5	36.1
Japan	38.9[a]	38.5	38.4	39.9
Netherlands	23.9[c]	23.4	26.6	37.3
Norway	30.1	27.1	36.4	44.3
Sweden	27.8	26.4	41.0	48.0
Switzerland	33.9	29.7	34.3	37.3
UK	29.0[b]	30.8	36.6	41.4
USA	21.2	28.8	38.0	44.1
Average	29.6	30.4	36.3	41.9

[a] 1913. [b] 1911. [c] 1909.

Sources: First column from P. Bairoch and Associates, *The Working Population and its Structure*, Université Libre de Bruxelles, 1968. Other columns from OECD, *Labour Force Statistics*, various editions. Netherlands 1973 and 1987 supplied by US Bureau of Labor Statistics.

TABLE C.3. *Labour Force Participation Rates around 1950*

	Year	Persons under 15	Persons 65 and over	Males 15–64	Females 15–64
Australia	1954	0.7	17.4	94.3	29.5
Austria	1951	n.a.	21.0	92.6[a]	47.8[a]
Belgium	1947	1.5	14.1	87.7	26.6
Canada	1951	0.2	21.9	89.5	26.5
Denmark	1950	2.4	21.5	94.8	47.3
Finland	1950	0.5	34.5	93.2	57.4
France	1954	1.0	22.6	89.6	43.6
Germany	1950	1.7	17.2	92.1	43.9
Italy	1957	n.a.	16.9	89.3[b]	29.7[b]
Japan	1955	n.a.	35.9	84.5[b]	51.8[b]
Netherlands	1947	n.a.	20.1	90.4[b]	28.9[b]
Norway	1950	0.0	23.6	93.6	28.9
Sweden	1950	0.2	21.0	93.3	33.7
Switzerland	1950	0.3	28.6	93.6	37.2
UK	1951	0.1	16.0	93.9	40.1
USA	1950	0.9	24.7	88.0	36.1
Average		0.8	22.3	91.3	38.1

[a] Numerator and denominator refer to age group 18–64.
[b] Numerator and denominator refer to age group 14–64.

have risen sharply and activity of young and old people has dropped (see Tables C.2, C.3, C.4).

The most drastic fall has been for the child population. In 1870 it was still quite common for children to work. Schooling was by no means universal or even compulsory. In most countries before the First World War, the age cut-off point used in censuses for inclusion in the working population was 10 years. Germany was exceptional in having 14 as the cut-off point as early as 1895. Between the two wars the cut-off point was generally raised to 14, and since 1960 it has been raised to 15 or 16.

Activity rates for older people have dropped substantially because of greater coverage and higher benefits of pension schemes and social security income guarantees, and the greater influence of regulations that prevent people from continuing at work (particularly as there has been a decline in the proportion of self-employed, who have greater

Table C.4. *Labour Force Participation Rates in 1987*

	Persons under 15	Persons 65 and over	Males 15–64	Females 15–64
Australia		4.8	85.0	57.5
Austria		3.1[a]	81.0	53.1
Belgium		4.1[a]	75.2	52.0
Canada		7.0	86.3	65.4
Denmark		12.4[a]	87.2	75.9
Finland		5.7	81.4	72.9
France		3.0	75.6	55.7
Germany		3.1	79.6[b]	51.3[b]
Italy		4.9	79.0	43.4
Japan		23.6	87.3	57.8
Netherlands		1.5	80.9	49.3
Norway		18.8	87.9	72.3
Sweden		8.5	83.9	79.4
Switzerland		18.9[a]	89.8	54.6
UK		4.8	88.3	62.6
USA		10.5	85.3	66.0
Average	0.0	8.4	83.4	60.6

[a] 1976. [b] 1986.

freedom to make their choices in this matter). The fall in activity rates for older people has been very substantial — big enough to offset the growth in the demographic importance of this group in most countries.

Within the age group between 15 and 64 years, from which the vast majority of the labour force is drawn, differing trends have been at work for males and females. For males of this age there has been a gradual but mild decrease in activity rates. Their activity was reduced by the increased availability of schooling and pensions. These forces also affected females, but were substantially outweighed by other influences which have generally raised female activity rates, particularly since 1950.

Aggregated information on labour supply is generally available only from decennial population censuses for the period before the Second World War. Statistical practice was not as well defined then as it is now. The concept of labour force was not yet developed, and the vaguer concept of 'active population' was generally used. This referred to the

TABLE C.5. *Structure of Employment in 1870, 1950, 1973, and 1987 (% of total employment)*

	1870			1950		
	Agriculture	Industry	Services	Agriculture	Industry	Services
Australia	30.0	33.0	37.0	14.6	36.5	48.9
Austria	65.0	19.2	15.8	34.0	35.4	30.6
Belgium	43.0	37.6	19.4	10.1	46.8	43.1
Canada	53.0	28.0	19.0	21.8	36.0	42.2
Denmark	51.7	n.a.	n.a.	25.1	33.3	41.6
Finland	65.0	16.3	18.7	46.0	27.7	26.3
France	49.2	27.8	23.0	28.3	34.9	36.8
Germany	49.5	28.7	21.8	22.2	43.0	34.8
Italy	62.0	23.0	15.0	45.4	28.6	26.0
Japan	70.1[a]	n.a.	n.a.	48.3	22.6	29.1
Netherlands	37.0	29.0	34.0	13.9	40.2	45.9
Norway	53.0	20.0	27.0	29.8	33.2	37.0
Sweden	54.0	21.0	25.0	20.3	40.8	38.9
Switzerland	49.8	n.a.	n.a.	16.5	46.4	37.1
UK	22.7	42.3	35.0	5.1	46.5	48.4
USA	50.0	24.4	25.6	13.0	33.3	53.7
Average	48.8	26.9	24.3	24.7	36.6	38.7

	1973			1987		
Australia	7.3	35.1	57.6	5.8	26.3	67.9
Austria	16.2	40.6	43.2	8.6	37.7	53.7
Belgium	3.8	39.5	56.7	2.6	27.7	69.7
Canada	6.5	30.4	63.1	4.8	25.2	70.0
Denmark	9.4	33.2	57.4	5.8	27.9	66.3
Finland	16.8	33.7	49.5	10.2	30.7	59.1
France	10.9	38.5	50.6	6.9	30.1	63.0
Germany	7.1	46.6	46.3	5.1	39.7	55.2
Italy	17.8	38.1	44.1	10.2	31.7	58.1
Japan	13.4	37.2	49.4	8.3	33.8	57.9
Netherlands	6.1	35.5	58.4	4.7	26.3	69.0
Norway	11.3	33.4	55.3	6.5	26.5	67.0
Sweden	7.1	36.8	56.1	3.9	29.8	66.3
Switzerland	7.5	44.9	47.6	6.4	37.7	55.9
UK	2.9	41.7	55.4	2.4	29.8	67.8
USA	4.1	32.3	63.6	3.0	26.6	70.4
Average	9.3	37.3	53.4	6.0	30.5	63.5

[a] 1872. [b] excludes Denmark, Japan, and Switzerland.

Sources: P. Bairoch and Associates, The Working Population and its Structure, Université Libre de Bruxelles, 1968; OECD, Labour Force Statistics, various issues; International Labour Organisation Yearbooks, and national sources. Agriculture includes agriculture, forestry, and fishing; industry includes mining, manufacturing, electricity, gas and water supply, and construction. Services is a residual including all other economic activity, private and governmental (including military).

normal rather than the actual occupation of the respondents and there-fore did not usually distinguish between the employed and unemployed. A major grey area was the treatment of female employment, particularly in agriculture.

Farming is largely an occupation of the self-employed and their families. The female labour force in agriculture is usually engaged in both house-hold and agricultural activities in most of the countries covered here, but statistical conventions have differed between countries. In the past, very low female participation rates were registered in Australian, British, Canadian, and US agriculture and very high rates in Austria, Denmark, and Japan. These large inter-country variations in female activity rates are due partly to differences in measurement convention. I have not made any direct adjustment to narrow the spread in female activity rates between countries, but have adjusted figures for agricultural females in cases where the female/male ratios in agriculture moved erratically between successive censuses in a particular country. The estimates presented for 1913–50 include adjustments of this kind for Austria, Denmark, France, Italy, and the USA.

For years before 1913 the problems of using census material become greater because variations between successive national census practice and between countries were bigger and the agricultural sector on which these problems were concentrated was larger then. For this period, it seemed preferable to assume that activity rates remained stable. Hence, for 1870–1913, I assumed that the labour force moved in the same proportion as the population of working age.

For the post-war period, the data situation has improved in terms of conceptual clarity, regularity, and quality of the information. All the countries cited here have in theory adopted the conventions recom-mended by the ILO for measuring labour force, employment, and un-employment, and these are also used by OECD. For 1950 onwards I therefore generally used the OECD labour force estimates. These estimates in absolute terms were linked to earlier years, using census measures of movements in activity rate (adjusted as described).

DETAILED SOURCE NOTES FOR LABOUR FORCE ESTIMATES

For *1870–1913*, it was assumed that, except for Australia and the UK, the labour force moved in step with the population of working age (i.e. those aged 15 to 64). Austria, Belgium, France, Germany, Italy, Netherlands, Norway, and Switzerland were derived by interpolation of census data as given in *The Aging of Population and its Economic and Social Implications*, United Nations, New York, 1956. Otherwise the following sources were used:

Australia: N. G. Butlin, 'Our 200 Years', *Queensland Calendar*, 1988.

Canada: M. C. Urquhart and K. A. H. Buckley, *Historical Statistics of Canada*, Cambridge, 1965, p. 16

Denmark: from S. A. Hansen, *Økonomisk vækst i Danmark*, vol. ii, Akademisk Forlag, Copenhagen, 1974, pp. 202–3.

Finland: *Statistical Yearbook of Finland*, 1973, Central Statistical Office, Helsinki.

Japan: M. Umemura and Associates, *Manpower*, vol. 2 of *Estimates of Long-Term Economic Statistics of Japan*, Toyo Keizai Shinposha, Tokyo, 1988.

Sweden: derived from *Historical Statistics of Sweden*, vol. i, *Population 1720–1950*, Central Bureau of Statistics, Stockholm, 1955, p. 22.

For *1913–50*, estimates of the change in the labour force were derived from A. Maddison, *Economic Growth in the West*, Allen & Unwin, 1964, Appendix D for Belgium, Denmark, Netherlands, Sweden, and Switzerland. Otherwise the following sources were used:

Australia: 1913–60 from M. Keating, 'Australian Work Force and Employment 1910–11 to 1960–61', *Australian Economic History Review*, Sept. 1967, adjusted to a calendar year basis.

Austria: 1913–53 from A. Kausel, *Oesterreichs Volkseinkommen 1913 bis 1963*, O. I. W., Vienna, 1965. I have adjusted Kausel's figure for 1913 downwards by 10 per cent. He apparently used census activity rates and the 1910 census showed more women than men employed in agriculture, whereas the 1920 census had just half as many women as men in agriculture. I have assumed that the 1913 female/male ratio in agriculture was two-thirds.

Canada: derived from Urquhart and Buckley, *Historical Statistics of Canada*, pp. 59–61. The 1913 age cut-off point was 10 for 1913, 14 thereafter. Figures are adjusted to offset the impact of the accession of Newfoundland in 1949.

Finland: derived from P. Bairoch, *The Working Population and Its Structure*, Brussels, 1968, p. 94.

France: 1929–38 from J.-J. Carré, P. Dubois, and E. Malinvaud, *La Croissance française*, Seuil, Paris, 1972, p. 80.

Germany: movement derived by applying activity rates (interpolated from census) to population of working age.

Italy: 1913–50 derived from O. Vitali, *Aspetti dello sviluppo economico italiano alla luce della ricostrozione della populazione attiva*, Istituto di Demografia, Rome, 1970, pp. 372–5.

Japan: 1913–38 from M. Umemura and Associates, *Manpower*, Toyo Keizai Shinposha, Tokyo, 1988. 1950–3, K. Ohkawa and H. Rosovsky, *Japanese Economic Growth*, Stanford, 1973, pp. 310–11 and 123 (refers to employment).

Norway: *Trends in Norwegian Economy 1865–1960*, CBS, Oslo, 1966, p. 29 (interpolated where necessary).

UK: 1870–1955 from C. H. Feinstein, *National Income Expenditure and Output of the United Kingdom 1855–1965*, Cambridge, 1972, pp. T125–7.

USA: 1910–50 derived from S. Lebergott, *Manpower in Economic Growth*, McGraw-Hill, New York, 1964, p. 512, adjusted to maintain a uniform female/male employment ratio in agriculture. 1870–1910 labour force assumed to move proportionally to population of working age (see *Historical Statistics of the United States: Colonial Times to 1970*, US Dept. of Commerce, Washington, DC, 1975, p. 15).

For *1950 onwards*, figures are taken from OECD, *Labour Force Statistics*, except as noted above for Australia (to 1960), Austria (to 1953), Japan (to 1953), the UK (to 1955), and as follows:

Austria: 1954–73 from *Statistisches Handbuch für die Bundesrepublik Österreich*.

Denmark: 1950–60 from Hansen, *Økonomisk vækst i Danmark*, pp. 203–4.

France: 1950 onwards from M. Granier, 'Séries longues démo-économiques', Aix-en-Provence, 1988, mimeographed.

Germany: from *Mitteilungen aus der Arbeitsmarkt und Berufsforschung*, figures include Saar and West Berlin throughout.

Italy: the official labour force sample survey has been affected by various problems which have not been satisfactorily settled. A major problem is that wage and working condition requirements, and the high level of social security payments, give both workers and employers an incentive to work without registering or reporting, either by working at home or by not declaring work in other work-places. When the national accounts were revised upwards for the bench-mark year 1982, the authorities estimated a total of 28 million 'jobs' compared with 22.2 million people figuring in the official labour force estimates: see *Conti economici nazionali, anni 1980–86*, ISTAT, Rome, 1987, pp. 9 and 31. I adjusted the official figures upwards to allow for this underground employment. My 1987 adjustment added 17.6 per cent to employment.

Japan: figures adjusted to include Okinawa throughout.

Netherlands: 1950–60 from Dutch Statistical Office, 1973 onwards supplied by US Bureau of Labor Statistics.

UK: 1950–60 from *British Labour Statistics Yearbook 1976*, HMSO, London, 1978.

USA: all years before 1960 adjusted upwards by 0.41 per cent to include Alaska and Hawaii.

UNEMPLOYMENT

Virtually all Western countries now publish an official measure of unemployment, but the scope of these different national indicators still varies a good deal, so that international comparison of unemployment rates can be made only with considerable reservation and after careful adjustment. The same is true of any long-period comparison because concepts and measures of unemployment have changed over time. The historical sequence has tended to be (1) figures for trade union members; (2) figures for those applying for jobs at unemployment offices; (3) figures for those claiming state insurance benefits (under schemes whose coverage has grown steadily and is now pretty universal for wage and salary-earners in most countries); (4) figures derived from census-type enquiries or labour force sample surveys.

Sample surveys are more comprehensive in coverage than other sources of data, because they cover the whole population, and ask questions from people who may have little incentive to register as unemployed, such as women and students seeking part-time jobs and new entrants to the labour market. However, not all countries treat the temporarily unemployed in the same way, and not all of them apply the same criteria for testing job search or availability for work.

Most of the countries considered here now have regular labour force sample surveys of the kind carried out in the USA since 1940, but the questionnaires differ between countries in ways that affect the count of the unemployed, and the classification of people as employed or unemployed, active or inactive, is influenced by differences in national tradition or labour market institutions which may have nothing to do with the degree of labour slack.

In spite of these pitfalls, there are a number of useful comparative studies of unemployment both in terms of international variation and historical trends. The ILO has performed a useful job in publishing unemployment statistics for most of the countries mentioned here since the 1920s (in the ILO *Yearbooks* issued from 1936 onwards), and although its own publications do not carry figures adjusted to ensure comparability, it has helped promote progress in this direction by meetings of the various Conferences of Labour Statisticians, which in 1954 reached agreement on standardized definitions (on US lines) of the labour force, employment, and unemployment. These ILO guidelines left some grey areas where incomparabilities exist, but OECD has made recommendations to clarify most of these problems (see *Measuring Employment and Unemployment*, OECD, Paris, 1979). A good historical supplement to the ILO material is W. Galenson and A. Zellner, 'International Comparison of Unemployment Rates', in *The Measurement and Behaviour of Unemployment*, NBER, Princeton, 1957. My own earlier work, *Economic Growth in the West*, Appendix E, contains estimates adjusted to improve international comparability of unemployment levels, and more recent estimates of this kind are available in C. Sorrentino, *International Comparisons of Unemployment*, US Bureau of Labor Statistics, Washington, DC, 1978.

For years prior to the First World War, the only countries for which there are reasonable series for any length of time are the UK and the USA: for the UK, a series for 1855–1914 in Feinstein, *National Income, Expenditure and Output of the UK*, pp. T125–6; and for the USA annual estimates are presented by S. Lebergott, *Manpower in Economic Growth*, McGraw-Hill, New York, 1964, p. 512.

For 1920–38, unemployment ratios for Belgium, Canada, Germany, Italy, Netherlands, Sweden, Switzerland, and the USA from sources indicated in Maddison, *Economic Growth in the West*. Otherwise from sources cited below.

Recently Barry Eichengreen and T. J. Hatton, *Interwar Unemployment in International Perspective*, Kluwer, Dordrecht, 1988, p. 10, have queried my estimates for 1920–38. They are reluctant to believe that unemployment was lower in Europe in the 1930s than in North America, and they suggest averaging my estimates with those for the much higher industrial unemployment. I regard their critique as invalid because they believe unemployment figures should include the underemployed, and their hybrid unemployment indicator is inconsistent with measures of employment and labour force.

Australia: 1920–60, from M. Keating, 'Australian Work Force and Employment 1910–11 to 1960–1', *Australian Economic History Review*, Sept. 1967, adjusted to a calendar year basis.

Austria: A. Kausel, N. Nemeth, and H. Seidel, 'Österreichs Volkseinkommen, 1913–63', *Monatsberichte des Österreichischen Instituts für Wirschaftsforschung*, 14. Sonderheft, Vienna, Aug. 1965, for years up to 1953. 1954 onwards from *Statistisches Handbuch für die Republik Österreich*, 1977, p. 74.

Denmark: 1920–70, Hansen, *Økonomisk vækst i Danmark*, pp. 203–4.

Finland: 1920–38 figures supplied by Kaarina Vattula of University of Helsinki. The figures refer to unfilled applications for work at labour exchanges.

France: 1920–50, J.-J. Carré, P. Dubois, and E. Malinvaud, *La Croissance française*, Seuil, Paris, 1972, pp. 80, 676–7.

Norway: O. Aukrust and J. Bjerke, 'Real Capital and Economic Growth in Norway, 1900–50', *Income and Wealth*, Series VIII, Bowes and Bowes, London, 1959, p. 116, and *Trends in Norwegian Economy 1865–1960*, CBS, Oslo, 1966, p. 29.

UK: Feinstein, *National Income, Expenditure and Output of the UK*, p. T126.

Except as specified, for 1950 onwards the figures were derived from OECD, *Labour Force Statistics*. France from R. Granier, 'Séries longues démo-économiques', Aix-en-Provence, 1988, mimeographed. Estimates for Netherlands for 1973 onwards from US Bureau of Labor Statistics, 'Comparative Labor Force Statistics 1959–89', May 1990. Germany from *Mitteilungen*, op. cit. An upward adjustment was made for Norway to compensate for partial coverage before 1972. For Sweden the 1950–60 figures are from Maddison, *Economic Growth in the West*.

EMPLOYMENT

Employment was assumed to move parallel to the labour force for 1870–1913. For 1913 onwards the figures were derived by subtracting unemployment from the labour force.

ANNUAL WORKING HOURS PER PERSON

Working hours data are among the weakest of those used here. Most of the regular estimates presently available cover only part of the labour force (usually industrial workers), and not all of them reflect changes in holidays and vacations. For the period before the First World War, estimates have to be based on very limited material. As there are broad similarities in the long-run trend of working hours, it was assumed for 1870–1913 that weekly hours per person were the same in all the countries, i.e. 60 in 1870 and 53.8 in 1913, with interpolations for intermediate years. The 53.8 figure for 1913 is derived from the *Hours and Earning Inquiry* for the United Kingdom in 1906 carried out by the Board of Trade. The figure includes overtime; see A. Maddison, 'Output, Employment and Productivity in British Manufacturing in the Last Half Century', *Bulletin of the Oxford University Institute of Statistics*, November 1955. The 1870 figure is derived from M. A. Bienefeld, *Working Hours in British Industry: An Economic History*, Weidenfeld & Nicolson, London, 1972, p. 111. Bienefeld's figures average 57.4 for 'normal' hours which were grossed up for overtime (using the 1906 ratio

— Bienefeld shows 1906 normal hours as 51 on p. 283). Differences between countries in my figures for this period reflect only differences in the length of holidays and vacations.

Weekly working hours for most countries were derived from Maddison, *Economic Growth in the West*, p. 228, for 1913–38; for Australia, Finland, and Japan, estimates for 1929 and 1938 were from ILO *Yearbooks of Labour Statistics*, 1938 and 1945–46 edns.; Germany, from *Statistisches Handbuch von Deutschland 1928–1944*, p. 480; for Austria it was assumed that working hours were the same as in Germany. For the USA, 1929 and 1938 were derived from J. W. Kendrick, *Productivity Trends in the United States*, Princeton, 1961, pp. 306–7 and 312–13, adjusted for days worked per year. For 1950 onwards, figures are from ILO *Yearbooks* unless otherwise stated.

Sources for hours per person from 1950 onwards are listed below. In all cases the figures are an average for both sexes. As average female hours are well below those of males, it is important that both be represented in the sample. *In the case of countries marked with an asterisk the source used made no allowance for changes in days worked per year.* In these cases, the figures were adjusted to exclude hours not worked as noted below.

Australia*: hours worked in agriculture, industry, and services are available from the quarterly labour force survey from 1965. Hours in 1950 and 1960 assumed to be the same as in the USA.

Austria: weekly hours actually worked by wage earners in manufacturing are available for 1964 onwards. This series was linked to the 1955–64 movement in weekly hours paid, and the 1950–5 movement in monthly hours paid.

Belgium: annual hours worked by the whole employed population (including the impact of holidays) for 1950–60 from S. Mendelbaum, 'Évolution de la quantité et de la durée du travail en Belgique de 1948 à 1962', *Cahiers économiques de Bruxelles*, no. 21, 1er trimestre, 1964, p. 87. 1961 onwards (Oct.) figures on weekly hours of '*ouvriers inscrits*' in manufacturing and construction from *Bulletin de statistique*, Institut National de Statistique.

Canada*: hours paid per week of wage-earners in manufacturing including hours spent on vacation or public holidays, sick leave, and other paid leave.

Denmark: total annual hours worked by wage-earners in industry, excluding hours lost in industrial disputes, holidays, sick-days, and compensatory leave were supplied by the Danish Statistical Office.

Finland*: 1950–73 weekly hours worked by wage-earners in industry. Data were averages for February, May, September, and November.

1987 figures derived from *National Accounts 1979–1988*, Helsinki (these are adjusted for holidays and sickness absence).

France, Germany, and UK: estimated as shown in A. Maddison, 'Monitoring the Labour Market', *Review of Income and Wealth*, June 1980.

Italy: monthly hours actually worked by wage-earners in industry and construction multiplied by 12. Ministry of Labour quarterly survey figures were used as given in *Rassegna di Statistiche del Lavoro*. Alternative figures are available for recent years in the labour force sample survey which shows considerably longer hours. However, the latter source covers only four weeks in the year and excludes many 'occasional workers'.

Japan: monthly hours worked by regular employees in establishments with 5 or more employees multiplied by 12, *Yearbooks of Labour Statistics*, Tokyo, and *50 Years History of Monthly Labour Statistics*, Ministry of Labour, Tokyo, 1974 (in Japanese). As the labour survey results show shorter hours worked by self-employed and temporary workers, the figures for regular workers for 1950 onwards were reduced by 5.4 per cent.

Netherlands: detailed estimates were made from national sources in the same way as for France, Germany, and UK.

Norway: hours worked per calendar week by wage-earners in manufacturing and mining. This series includes the impact of vacations, holdidays, etc. The figures published in ILO *Yearbooks* (and in *Lonnstatistikk*) are shown separately for males and females. These were averaged by weighting males 80 per cent and females 20 per cent (their respective shares in employment in manufacturing and mining as derived from OECD, *Labour Force Statistics*). 1950 estimate made from earlier series on weekly hours, reduced by 9 per cent to eliminate vacations and public holidays.

Sweden: for years since 1960, the national accounts publications show total hours worked by entrepreneurs and employees in the whole economy. The figures refer to actual time worked and reflect the impact of sickness absence, public holidays, vacations, strikes, etc. 1950–60 extrapolated backwards by using movement in monthly hours worked in industry (ILO *Yearbooks*).

*Switzerland**: weekly hours paid in manufacturing per 'production and related worker'. 1973 onwards, figures adjusted downwards by 1.34 per cent to achieve continuity with earlier series. The figures from 1950 onwards were reduced by 5 per cent throughout, as they referred only to manual workers.

UK: as for France.

USA*: average weekly hours of production or non-supervisory workers on private non-agricultural payrolls, from *Employment and Earnings*, various issues.

Adjustment for Holidays and Vacations

Where necessary, the figures were adjusted to take account of changes in public holidays and vacations. The sources used were those cited in Maddison, *Economic Growth in the West*, 'Wages Policy at Home and Abroad', *Westminster Bank Review*, Nov. 1962, p. 33; *New Patterns for Working Time*, OECD, 1973; E. F. Denison, *Why Growth Rates Differ*, Brookings Institution, Washington, DC, 1962, p. 363; A. A. Evans, *Hours of Work in Industrialised Countries*, ILO, Geneva, 1975; and J. E. Buckley, 'Variations in Holidays, Vacations and Area Pay Levels', *Monthly Labor Review*, Feb. 1989.

Adjustment for Sickness

It is important to adjust labour input to exclude time lost through sickness, which is higher in Europe than in the USA as benefits are more generous. For France, Germany, and the UK see sources cited in Maddison, 'Monitoring the Labour Market'. For the Netherlands see sources used in A. Maddison and B. S. Wilpstra, *Unemployment: The European Perspective*, Croom Helm, London, 1982, pp. 80–1. For the USA, for 1950 onwards the time loss was assumed to be 3.5 per cent: see J. N. Hedges, 'Absence from Work — Measuring the Hours Lost', *Monthly Labor Review*, Oct. 1977. For Belgium, Denmark, Italy, Japan, Norway, and Sweden the basic estimate of hours worked for the post-war period seemed to exclude time lost on sickness. For Canada the loss was assumed to be the same as in the USA (3.5 per cent). In Australia, Austria, Finland, and Switzerland, a 5 per cent working-time loss was assumed from this cause for 1950 onwards. For all countries it was assumed that the time loss from this cause was 2.5 per cent for the period 1870–1938.

PRODUCTIVITY LEVELS

Table C.11 is derived by dividing the estimates of GDP in Table A.2 by the estimates of total working hours in Table C.10. Both of these are adjusted to exclude the impact of changes in geographic coverage. A comparison of Tables A.2 and A.3 shows the impact of these geographic changes on GDP. A comparison of Table B.7 with Tables B.2 to B.4 shows their impact on demography.

In the case of the UK, the productivity level before 1920 was lower than indicated in Table C.11, because Irish productivity was lower than that in the rest of the UK. The estimates in Table C.12 show the derivation of the UK productivity levels, including the whole of Ireland, for 1700–1913.

The rough estimates of Dutch productivity (Table C.13) are derived by assuming simply that the Dutch employment population ratio in 1700–85 was the same as that of the UK in 1785 and that working hours per person employed were 3,000 per year. The sources for Dutch GDP per head and population estimates are described in Appendices A and B.

TABLE C.6. *(a) Unemployment as a Percentage of the Total Labour Force, 1920–1938*

	Australia	Austria	Belgium	Canada	Denmark	Finland	France	Germany
1920	4.6				3.0	1.1		1.7
1921	5.9		6.1	5.8	10.0	1.8	2.7	1.2
1922	5.5		1.9	4.4	9.5	1.4		0.7
1923	4.9		0.6	3.2	6.5	1.0		4.5
1924	5.5	5.4	0.6	4.5	5.5	1.2		5.8
1925	5.6	6.3	0.9	4.4	7.5	2.0		3.0
1926	4.6	7.0	0.8	3.0	10.5	1.6	1.2	8.0
1927	5.2	6.2	1.1	1.8	11.0	1.5		3.9
1928	6.4	5.3	0.6	1.7	9.0	1.5		3.8
1929	8.2	5.5	0.8	2.9	8.0	2.8	1.2	5.9
1930	13.1	7.0	2.2	9.1	7.0	4.0		9.5
1931	17.9	9.7	6.8	11.6	9.0	4.6	2.2	13.9
1932	19.1	13.7	11.9	17.6	16.0	5.8		17.2
1933	17.4	16.3	10.6	19.3	14.5	6.2		14.8
1934	15.0	16.1	11.8	14.5	11.0	4.4		8.3
1935	12.5	15.2	11.1	14.2	10.0	3.7		6.5
1936	9.9	15.2	8.4	12.8	9.5	2.7	4.5	4.8
1937	8.1	13.7	7.2	9.4	11.0	2.6		2.7
1938	8.1	8.1	8.7	11.4	10.5	2.6	3.7	1.3

	Italy	Netherlands	Norway	Sweden	Switzerland	UK	USA
1920		1.7		1.3		1.9	3.9
1921		2.6	5.6	6.4		11.0	11.4
1922		3.2	5.2	5.5		9.6	7.2
1923		3.3	1.3	2.9		8.0	3.0
1924		2.6	0.3	2.4		7.1	5.3
1925		2.4	3.4	2.6		7.7	3.8
1926		2.1	10.4	2.9		8.6	1.9
1927		2.2	11.3	2.9		6.7	3.9
1928		1.6	7.6	2.4		7.4	4.3
1929	1.7	1.7	5.4	2.4	0.4	7.2	3.1
1930	2.5	2.3	6.2	3.3	0.7	11.1	8.7
1931	4.3	4.3	10.2	4.8	1.2	14.8	15.8
1932	5.8	8.3	9.5	6.8	2.8	15.3	23.5
1933	5.9	9.7	9.7	7.3	3.5	13.9	24.7
1934	5.6	9.8	9.4	6.4	3.3	11.7	21.6
1935		11.2	8.7	6.2	4.2	10.8	20.0
1936		11.9	7.2	5.3	4.7	9.2	16.8
1937	5.0	10.5	6.0	5.1	3.6	7.7	14.2
1938	4.6	9.9	5.8	5.1	3.3	9.2	18.8

TABLE C.6. (b) Unemployment as a Percentage of the Total Labour Force, 1950–1989

	Australia	Austria	Belgium	Canada	Denmark	Finland	France	Germany
1950	1.5	3.9	5.0	3.6	4.0	1.0	2.3	8.2
1951	1.3	3.5	4.4	2.4	4.6	0.3	2.1	7.3
1952	2.2	4.7	5.1	2.9	5.8	0.4	2.1	7.0
1953	2.5	5.5	5.3	2.9	4.4	1.5	2.6	6.2
1954	1.7	5.0	5.0	4.5	3.8	1.0	2.8	5.6
1955	1.4	3.6	3.9	4.3	4.5	0.4	2.4	4.3
1956	1.8	3.4	2.8	3.3	5.1	2.2	1.8	3.4
1957	2.3	3.2	2.3	4.5	4.9	2.2	1.4	2.9
1958	2.7	3.4	3.3	6.9	4.5	2.2	1.6	3.0
1959	2.6	3.1	4.0	5.8	3.1	2.1	1.9	2.0
1960	2.5	2.3	3.3	6.8	2.1	1.4	1.8	1.0
1961	2.3	1.8	2.5	7.0	1.9	1.2	1.5	0.7
1962	2.2	1.9	2.1	5.8	1.6	1.3	1.4	0.6
1963	1.8	2.0	1.7	5.4	2.1	1.5	1.3	0.7
1964	1.6	1.9	1.3	4.3	1.2	1.5	1.1	0.6
1965	1.5	1.9	1.5	3.6	1.0	1.4	1.3	0.5
1966	1.7	1.7	1.6	3.3	1.1	1.5	1.4	0.6
1967	1.8	1.8	2.3	3.8	1.2	2.9	1.8	1.7
1968	1.7	1.6	2.7	4.4	1.6	3.9	2.1	1.2

1969	1.7	2.0	2.1	4.4	1.1	2.8	2.3	0.7
1970	1.6	1.4	1.8	5.6	0.7	1.9	2.4	0.6
1971	1.8	1.3	1.7	6.1	1.1	2.2	2.6	0.7
1972	2.6	1.2	2.7	6.2	0.9	2.5	2.8	0.8
1973	2.3	1.1	2.7	5.5	0.9	2.3	2.7	0.8
1974	2.6	1.3	3.0	5.3	3.5	1.7	2.8	1.6
1975	4.8	1.8	5.0	6.9	4.9	2.2	4.0	3.6
1976	4.7	1.8	6.4	7.1	6.3	3.8	4.4	3.7
1977	5.6	1.6	7.4	8.0	7.3	5.8	4.9	3.6
1978	6.2	2.1	7.9	8.3	8.3	7.2	5.2	3.5
1979	6.2	2.1	8.2	7.4	6.0	5.9	5.9	3.2
1980	6.0	1.9	8.8	7.4	6.9	4.6	6.3	3.0
1981	5.7	2.5	10.8	7.5	10.3	4.8	7.4	4.4
1982	7.1	3.5	12.6	10.9	11.0	5.3	8.1	6.1
1983	9.9	4.1	12.1	11.8	11.4	5.4	8.3	8.0
1984	8.9	3.8	12.1	11.2	8.5	5.2	9.7	7.1
1985	8.2	3.6	11.3	10.4	7.3	5.0	10.2	7.2
1986	8.0	3.1	11.2	9.5	5.5	5.3	10.4	6.4
1987	8.0	3.8	11.0	8.8	6.9	5.0	10.5	6.2
1988	7.2	3.6	9.7	7.7	7.2	4.5	10.0	6.2
1989	6.1	(3.6)	8.1	7.5	(7.2)	3.4	9.4	5.6

Note: Bracketed figures are estimates.

TABLE C.6. (b) (Cont.) Unemployment as a Percentage of the Total Labour Force, 1950–1989

	Italy	Japan	Netherlands	Norway	Sweden	Switzerland	UK	USA
1950	6.9	1.9	2.8	1.2	1.7	0.0	2.5	5.2
1951	7.3	1.7	3.2	1.5	1.6	0.0	2.2	3.2
1952	7.8	1.9	4.9	1.6	1.7	0.0	3.0	2.9
1953	8.1	1.7	3.5	1.9	1.9	0.0	2.6	2.8
1954	8.3	2.2	2.3	1.8	1.8	0.0	2.3	5.3
1955	7.0	2.5	1.5	1.6	1.8	0.0	2.1	4.2
1956	8.7	2.3	1.0	1.9	1.6	0.0	2.2	4.0
1957	7.0	1.9	1.5	2.1	1.7	0.0	2.4	4.2
1958	6.0	2.0	3.0	3.3	2.0	0.0	3.0	6.6
1959	5.2	2.2	2.1	3.2	1.8	0.0	3.0	5.3
1960	3.9	1.7	1.2	2.3	1.7	0.0	2.2	5.4
1961	3.4	1.4	0.9	1.8	1.5	0.0	2.0	6.5
1962	2.9	1.3	0.9	2.0	1.5	0.0	2.8	5.4
1963	2.5	1.2	0.9	2.4	1.7	0.0	3.4	5.5
1964	3.9	1.2	0.8	2.1	1.6	0.0	2.5	5.0
1965	5.0	1.1	1.0	1.7	1.2	0.0	2.2	4.4
1966	5.4	1.3	1.4	1.6	1.6	0.0	2.3	3.7
1967	5.1	1.3	2.8	1.5	2.1	0.0	3.4	3.8
1968	5.3	1.2	2.5	2.2	2.2	0.0	3.3	3.5

1969	5.2	1.1	1.8	2.1	1.9	0.0	3.0	3.5
1970	4.9	1.1	1.6	1.5	1.5	0.0	3.1	4.9
1971	4.9	1.2	2.3	1.5	2.5	0.0	3.8	5.9
1972	6.3	1.4	3.9	1.7	2.7	0.0	4.0	5.5
1973	6.2	1.3	3.9	1.5	2.5	0.0	3.0	4.8
1974	5.3	1.4	4.4	1.5	2.0	0.0	2.9	5.5
1975	5.8	1.9	5.9	2.3	1.6	0.4	4.3	8.3
1976	6.6	2.0	6.3	1.8	1.6	0.7	5.6	7.6
1977	7.0	2.0	6.0	1.5	1.8	0.4	6.0	6.9
1978	7.1	2.2	6.2	1.8	2.2	0.3	5.9	6.0
1979	7.6	2.1	6.6	2.0	2.1	0.3	5.0	5.8
1980	7.5	2.0	6.0	1.6	2.0	0.2	6.4	7.0
1981	7.8	2.2	8.5	2.0	2.5	0.2	9.8	7.5
1982	8.4	2.4	11.4	2.6	3.2	0.4	11.3	9.5
1983	8.8	2.6	12.0	3.4	3.5	0.9	12.4	9.5
1984	9.4	2.7	11.8	3.1	3.1	1.1	11.7	7.4
1985	9.6	2.6	10.6	2.6	2.8	0.9	11.2	7.1
1986	10.5	2.8	9.9	2.0	2.7	0.8	11.2	6.9
1987	10.9	2.8	9.6	2.1	1.9	0.7	10.3	6.1
1988	11.0	2.5	9.2	3.2	1.6	0.6	8.5	5.4
1989	10.9	2.3	8.3	4.9	1.4	0.5	7.1	5.2

TABLE C.7. *Total Labour Force, 1870–1989* (mid-year, '000s)

	Australia	Austria	Belgium	Canada	Denmark	Finland	France	Germany
1870	651	2,119	2,209	1,305	872	794	18,164	10,459
1890	1,617	2,479	2,602	1,855	1,013	1,058	18,929	12,276
1913	2,046	3,186	3,484	3,107	1,358	1,338	19,802	17,638
1929	2,568	3,474	3,665	4,079	1,604	1,702	20,540	20,231
1938	2,821	3,389	3,633	4,722	1,943	1,968	19,490	21,483
1950	3,510	3,345	3,515	5,216	2,060	1,978	20,070	23,053
1960	4,170	3,364	3,573	6,530	2,199	2,152	20,055	26,351
1973	5,962	3,202	3,840	9,358	2,447	2,245	22,289	27,339
1987	7,796	3,427	4,115	13,090	2,879	2,583	24,294	29,506
1988	8,002	3,430	4,127	13,354	2,903	2,574	24,343	29,673
1989	8,303	3,450		13,532		2,583	24,472	29,841

	Italy	Japan	Netherlands	Norway	Sweden	Switzerland	UK	USA
1870	14,196	19,060	1,403	717	1,945	1,298	12,751	15,352
1890	15,867	20,771	1,705	795	2,179	1,431	15,075	26,043
1913	18,190	26,266	2,365	999	2,631	1,923	18,964	40,489
1929	19,345	29,897	3,075	1,196	3,224	2,004	20,405	49,471
1938	20,217	32,906	3,517	1,345	3,329	2,052	22,927	55,350
1950	20,274	36,374	4,239	1,446	3,481	2,237	22,965	65,016
1960	21,914	45,426	4,686	1,457	3,679	2,706	24,777	73,126
1973	24,209	53,260	5,310	1,702	3,977	3,277	25,633	91,756
1987	28,265	60,840	6,640	2,171	4,421	3,463	27,979	121,602
1988	28,669	61,660	6,660	2,183	4,471	3,500	28,256	123,378
1989	28,527	62,700	6,750	2,155	4,527	3,535	28,499	125,557

Note: Estimates have been adjusted to exclude the impact of frontier changes. The figures therefore refer throughout to the 1989 boundaries.

TABLE C.8. *Total Employment, 1870–1989 (mid-year, '000s)*

	Australia	Austria	Belgium	Canada	Denmark	Finland	France	Germany
1870	630	2,077	2,141	1,266	820	785	17,800	10,260
1890	1,563	2,429	2,521	1,799	952	1,046	18,550	12,043
1913	1,943	3,122	3,376	3,014	1,277	1,323	19,406	17,303
1929	2,355	3,282	3,636	3,960	1,476	1,654	20,170	19,037
1938	2,592	3,113	3,316	4,183	1,739	1,917	18,769	21,204
1950	3,459	3,215	3,341	5,030	1,978	1,959	19,663	21,164
1960	4,065	3,285	3,456	6,084	2,152	2,121	19,709	26,080
1973	5,852	3,160	3,831	8,843	2,426	2,194	21,696	27,066
1987	7,170	3,261	3,649	11,940	2,643	2,453	21,748	27,050
1988	7,436	3,271	3,702	12,323	2,695	2,458	21,920	27,264
1989	7,795	3,301		12,564		2,494	22,172	27,635

	Italy	Japan	Netherlands	Norway	Sweden	Switzerland	UK	USA
1870	13,770	18,684	1,382	706	1,923	1,285	12,285	14,720
1890	15,391	20,305	1,680	783	2,155	1,416	14,764	24,970
1913	17,644	25,751	2,330	984	2,602	1,904	18,566	38,821
1929	19,016	29,332	3,023	1,132	3,146	1,995	18,936	47,915
1938	19,287	32,290	3,169	1,267	3,159	1,984	20,818	44,917
1950	18,875	35,683	4,120	1,428	3,422	2,237	22,400	61,651
1960	21,059	44,670	4,630	1,423	3,616	2,706	24,225	69,195
1973	22,708	52,590	5,150	1,676	3,879	3,277	25,076	87,391
1987	24,930	59,110	5,990	2,126	4,337	3,438	25,074	114,177
1988	25,276	60,110	6,040	2,114	4,399	3,478	25,915	116,677
1989	25,156	61,280	6,150	2,049	4,465	3,518	26,756	119,030

Note: Estimates have been adjusted to exclude the impact of frontier changes. The figures therefore refer throughout to the 1989 boundaries.

TABLE C.9. *Annual Hours Worked per Person, 1870–1989*

	Australia	Austria	Belgium	Canada	Denmark	Finland	France	Germany
1870	2,945	2,935	2,964	2,964	2,945	2,945	2,945	2,941
1890	2,770	2,760	2,789	2,789	2,770	2,770	2,770	2,765
1913	2,588	2,580	2,605	2,605	2,588	2,588	2,588	2,584
1929	2,139	2,281	2,272	2,399	2,279	2,123	2,297	2,284
1938	2,110	2,312	2,267	2,240	2,267	2,183	1,848	2,316
1950	1,838	1,976	2,283	1,967	2,283	2,035	1,926	2,316
1960	1,767	1,951	2,174	1,877	2,127	2,041	1,919	2,081
1973	1,708	1,778	1,872	1,788	1,742	1,707	1,771	1,804
1987	1,631	1,595	1,620	1,673	1,669	1,663	1,543	1,620
1988		1,607			1,654	1,673		1,623
1989		1,591			1,654	1,655		1,607

	Italy	Japan	Netherlands	Norway	Sweden	Switzerland	UK	USA
1870	2,886	2,945	2,964	2,945	2,945	2,984	2,984	2,964
1890	2,714	2,770	2,789	2,770	2,770	2,807	2,807	2,789
1913	2,536	2,588	2,605	2,588	2,588	2,624	2,624	2,605
1929	2,228	2,364	2,260	2,283	2,283	2,340	2,286	2,342
1938	1,927	2,391	2,244	2,128	2,204	2,257	2,267	2,062
1950	1,997	2,166	2,208	2,101	1,951	2,144	1,958	1,867
1960	2,059	2,318	2,051	1,997	1,823	2,065	1,913	1,795
1973	1,612	2,093	1,751	1,721	1,571	1,930	1,688	1,717
1987	1,528[a]	2,020	1,387	1,486	1,466	1,794	1,557	1,608
1988		2,020						1,604
1989		1,998					1,552	1,604

[a] 1985.

TABLE C.10. *Total Hours Worked per Year, 1870–1989 (million hours)*

	Australia	Austria	Belgium	Canada	Denmark	Finland	France	Germany
1870	1,855	6,096	6,346	3,752	2,415	2,312	52,421	30,175
1890	4,330	6,704	7,031	5,017	2,637	2,897	51,384	33,299
1913	5,029	8,055	8,795	7,852	3,305	3,424	50,223	44,711
1929	5,037	7,486	8,261	9,500	3,364	3,511	46,330	43,481
1938	5,469	7,197	7,517	9,370	3,942	4,149	34,685	49,109
1950	6,358	6,353	7,628	9,894	4,516	3,987	37,871	49,016
1960	7,183	6,409	7,513	11,420	4,577	4,329	37,822	54,273
1973	9,995	5,619	7,172	15,811	4,226	3,745	38,424	48,827
1987	11,694	5,201	5,911	19,976	4,411	4,080	33,557	43,821
1988		5,256			4,458	4,112		44,249
1989		5,252				4,127		44,409

	Italy	Japan	Netherlands	Norway	Sweden	Switzerland	UK	USA
1870	39,740	55,024	4,096	2,079	5,663	3,834	36,658	43,630
1890	41,771	56,245	4,686	2,169	5,969	3,975	41,443	69,641
1913	44,745	66,644	6,070	2,547	6,734	4,996	48,717	101,129
1929	42,368	69,341	6,832	2,584	7,182	4,668	43,288	112,217
1938	37,166	77,205	7,111	2,696	6,962	4,478	47,194	92,619
1950	37,693	77,289	9,097	3,000	6,676	4,796	43,859	115,102
1960	43,361	103,545	9,496	2,842	6,592	5,588	46,342	124,205
1973	36,605	110,071	9,018	2,884	6,094	6,325	42,328	150,050
1987	38,093	119,402	8,308	3,159	6,358	6,168	39,040	183,597
1988		121,422						187,150
1989		122,437					41,525	190,924

TABLE C.11. *Productivity 1870–1989: GDP per Man-Hour in 1985 US Relative Prices ($)*

	Australia	Austria	Belgium	Canada	Denmark	Finland	France	Germany
1870	2.73	1.06	1.68	1.32	1.21	0.71	1.15	1.04
1890	2.82	1.51	2.30	1.81	1.69	0.92	1.52	1.52
1913	4.34	2.24	2.85	3.52	2.72	1.53	2.26	2.32
1929	5.32	2.53	3.80	4.52	4.09	2.17	3.30	2.89
1938	5.88	2.57	4.17	4.57	4.25	2.57	4.25	3.57
1950	7.63	3.11	4.79	8.49	4.94	3.50	4.58	3.40
1960	10.04	5.52	6.53	11.51	6.66	5.22	7.17	6.62
1973	13.96	11.67	12.79	16.56	12.51	11.30	14.00	12.83
1987	17.93	17.05	19.86	21.15	15.57	15.35	21.63	18.35
1988		17.57			15.34	16.02		18.85
1989		18.25				16.75		19.53

	Italy	Japan	Netherlands	Norway	Sweden	Switzerland	UK	USA
1870	0.85	0.39	1.82	0.99	0.97	1.28	2.15[b]	2.06
1890	0.99	0.58	2.49	1.36	1.32	1.74	2.86[b]	2.82
1913	1.72	0.86	3.23	2.00	2.04	2.39	3.63[b]	4.68
1929	2.38	1.48	5.09	3.12	2.61	3.95	4.58	6.88
1938	3.12	1.83	5.03	3.93	3.38	4.33	4.97	7.81
1950	3.52	1.69	5.23	4.94	5.60	6.42	6.49	11.39
1960	5.52	2.94	7.87	7.55	7.80	8.65	8.15	14.54
1973	12.82	9.12	15.30	12.84	15.08	13.41	13.36	19.92
1987	18.25[a]	14.04	21.27	20.65	18.94	15.78	18.46	23.04
1988		14.60						23.65
1989		15.18					18.55	23.87

[a] Assumes hours worked per person to be same as in 1985.

[b] The UK figures (like those for all other countries in the table) are adjusted to exclude the impact of territorial changes. Including Southern Ireland, the UK figures were 2.04 in 1870, 2.71 in 1890, and 3.45 in 1913, see Table C12.

TABLE C.12. *Derivation of UK Productivity Levels, 1700–1913*

	GDP (1985 $ million)	Population ('000)	Employment ('000)	Annual hours worked per person	GDP per man-hour (1985 $)	GDP per head of population (1985 $)
1913	183,707	45,649	20,310	2,624	3.45	4,024
1890	122,900	37,485	16,150	2,807	2.71	3,279
1870	81,934	31,393	13,440	2,984	2.04	2,610
1820	29,834	21,240	8,665	3,000	1.15	1,405
1780	15,633	12,640	5,383	3,000	0.97	1,237
1700	8,652	8,400	3,717	3,000	0.78	1,030

Sources: This table includes the whole of Ireland. GDP and population figures from sources already described in Appendices A and B. Employment 1855–1913 from C. H. Feinstein, *National Income, Expenditure and Output of UK, 1855–1965*, Cambridge, 1972, pp. T125–6; earlier years derived from estimates of changes in population of working age.

TABLE C.13. *Derivation of Dutch Productivity Levels, 1580–1870*

	GDP	Population	Employment	Annual hours worked per person	GDP per man-hour	GDP per head of population (1985 $)
	(1985 $ million)	('000)	('000)		(1985 $)	
1870	7,463	3,615	1,382	2,964	1.82	2,064
1820	3,077	2,355	961	3,000	1.07	1,307
1700	2,877	1,900	842	3,000	1.14	1,514
1580	1,233	1,370	607	3,000	0.68	900

Sources: GDP and population figures from sources described in Appendices A and B. Employment assumed to move with population of working age.

APPENDIX D

NON-RESIDENTIAL REPRODUCIBLE
TANGIBLE FIXED CAPITAL STOCK

Estimates of capital stock can be made in two ways: (1) by wealth surveys, insurance valuations, company book-keeping, or stock exchange values; (2) by cumulating historical series on past investment and deducting assets which are scrapped, written off, or destroyed by war. The second (perpetual inventory) method is preferable because it produces figures whose meaning is clearer because all the hypotheses and calculations are transparent and consistent. It is now generally used in official estimates, though the Japanese statisticians use a post-war wealth survey bench-mark.

COVERAGE OF NON-RESIDENTIAL CAPITAL

The standard definition includes all non-residential structures, machinery and equipment, and vehicles. It excludes land and natural resources, intangibles like human capital or the stock of knowledge, precious metals, international monetary reserves, foreign assets, inventories, consumer durables, housing, and military items. The official estimates for Germany, the UK, and the USA conform to the desired coverage but those for France and Japan exclude all government assets.

'GROSS' AND 'NET'

Estimates are usually made both on a *net basis*, with allowance each year for 'depreciation' of old assets (i.e. for assets that are retired from use and for declines in the use value of existing *assets* that are not retired), and on a *gross basis*, where allowance is made only for retirement and not for the decline in use value. The gross concept is equivalent to assuming that all existing assets are as good as new; it is appropriate for measuring factor productivity and assessing production potential, because most assets in use are repaired and maintained in such a way that their productive capacity remains near to their original level throughout their life. Net values are useful in measuring profitability or rates of return, because they involve a discount for differences in the expected future life of assets.

The level of the net stock will always be lower than the gross, and the relationship between the growth of the two measures will depend on the past history of capital formation. The pace is similar when growth has proceeded steadily for long periods, but when investment accelerates, as in 1950–73, the net stock rises more rapidly than the gross. The converse is true if investment decelerates, as it did after 1973.

ASSET LIVES

The most difficult problem arises from the general ignorance about the actual length of life of assets. Assumptions about asset lives are always stylized and in substantial degree hypothetical, and they differ significantly between the countries covered here. Table D.1 provides a rough indicator of the spread in lives assumed in the official figures in four countries.

TABLE D.1. *Average Life Expectation of Non-Residential Fixed Capital Assets in Official Estimates*

	1950	1987
France	22.8[a]	16.0[a]
Germany	40.0	31.2
UK	38.6	28.9
USA	27.4	23.6[b]

[a] Private sector only. [b] 1985.

Note: Average life expectation is calculated here by dividing the end year gross stock by depreciation in the same year. If straight-line depreciation is used, this will provide a reasonable estimate of the average remaining life expectancy of all assets in the stock. The life expectancy of new assets just entering the stock (which is what we are discussing in the text) will be higher than the remaining life expectancies of existing assets. But in the absence of a handy measure of average lives of new assets in official statistics, this table gives a rough measure of intercountry variance.

It is clear that asset lives are assumed to be considerably shorter in France and the USA than in the other countries. Average expectation will of course depend on the mix of assets in the stock as well as on the lives assumed for particular items. Thus, the change in average lives that is general between 1950 and 1987 is due in some degree to a change in the asset mix.

In Germany and the UK, official statisticians assume that lives of individual assets have shortened gradually or stepwise over time. In these two countries this shortening means that the capital stock increases more slowly than if fixed lives were assumed. There is no strong ground for the German and British assumption that asset lives (within a particular category and excluding compositional effects) decline over time. Studies of second-hand markets do not suggest this and there is no real evidence of an accelerating pace of technical progress.

RETIREMENT PATTERNS USED TO CALCULATE OFFICIAL GROSS STOCKS

The simplest assumption about asset lives is that all goods of the same kind bought in the same year are scrapped together when their expected life is reached. This (rectangular) assumption makes no allowance for accidents, fires, etc., so a number of alternative dispersion patterns have been developed. The following retirement patterns are those predominantly used in the official statistics of the following five countries:

France   : lognormal
Germany : gamma probability density function
Japan   : rectangular
UK   : even spread 20 per cent on each side of average life
USA   : bell-wise spread 55 per cent on each side of average life.

DEPRECIATION FORMULAE (USED TO CALCULATE NET STOCK)

Most countries making official estimates of the net capital stock appear to use straight-line depreciation as the standard technique, but Germany makes no depreciation allowance for government infrastructure. Japan uses declining balance depreciation in its official capital stock measure (and survey data on actual depreciation practice for its national accounts!).

STANDARDIZED ESTIMATES

As the official estimates involve important differences of assumption of various kinds and particularly about asset lives, I have constructed standardized estimates for 6 countries using identical assumptions about lives, retirement, and depreciation patterns. They are presented in Tables D.5 to D.10 below. As a major purpose of this study is to measure distance between the follower countries and the lead country, the USA, the figures are converted into dollars at 1985 US relative prices using the same type of PPP conversion I used for GDP in Appendix A from data supplied by Eurostat. I also used asset lives which approximate as closely as possible to those for the USA. I distinguish two categories of asset, machinery and equipment with a 15-year life, and non-residential structures with a life of 40 years. I assume a rectangular retirement pattern and straight-line depreciation.

Thus I arrive at the average asset life expectations in Table D.2, calculated by the same method as in Table D.1. In the USA the results of Table D.2 and D.1 are very close but they are very different for the other countries.

In Table D.2 the inter-country disparity in lives is much smaller than in Table D.1 and is due essentially to differences in the distribution of assets between our two major categories. In France, Germany, the Netherlands, and the USA average lives fell between 1950 and 1987 because of the increased share of shorter-lived assets (see Table D.3). In Japan and the UK average lives increased for the opposite reason.

Table D.4 compares our results with the official estimates for Germany, the UK, and the USA. Such a comparison with the Netherlands is not possible as there are no official estimates, and for France and Japan it would be misleading because the official figures exclude all publicly owned assets.

TABLE D.2. *Average Life Expectation of Total Non-Residential Fixed Assets in Standardized Estimates Valued at 1985 US $*

	1950	1987
France	29.1	22.2
Germany	28.2	24.9
Japan	22.9	25.9
Netherlands	30.3	24.8
UK	20.1	21.2
USA	26.9	23.9

TABLE D.3. *Shares of Machinery and Equipment in Standardized Estimates of Gross Fixed Capital Stock Valued in 1985 US $*

	1950	1973	1987
France	18.9	43.3	43.8
Germany	22.4	34.6	31.5
Japan	41.8	36.9	29.9
Netherlands	15.8	32.0	32.8
UK	52.7	52.8	48.2
USA	26.1	30.5	36.2

It is clear from Table D.4 that the international comparability of official stock estimates is badly compromised by differences in the length of life assumptions (and to a lesser extent by the other differences mentioned above). Thus in 1950 the German and UK official capital stock level was much higher than the stock re-estimated with US lives. The differences in level narrow over time, because both Germany and the UK assume declining lives of assets. A consequence of this is that the official British and German estimates show slower growth rates for 1950–73 than the standardized figures.

The official estimates are more finely disaggregated than mine. Germany has 207 different types of asset — see H. Lützel, 'Estimates of Capital Stock by Industries in the Federal Republic of Germany', *Review of Income and Wealth*, March 1977, p. 65. Average lives for the non-residential assets of enterprises are 14 years for machinery and equipment and 57 for structures: see L. Schmidt, 'Reproduzierbares Anlagevermögen', *Wirtschaft und Statistik*, July 1986, p. 503. The UK has four types of non-residential asset, whose lives vary across a 36-industry division — see CSO, *UK National Accounts: Sources and Methods*, HMSO, London, 1985, p. 200. The impact of compositional changes will thus be different from mine. However, this is not likely to be a major reason for differences between the standardized and the official estimates, because my crude two-way asset breakdown replicates the US level very well although the official figures are disaggregated into 95 different types of non-residential asset.

Another possible difference is in the investment series used in cumulating the stock of assets. For the USA, I used the same annual asset

TABLE D.4. *Confrontation of Official Estimates of Total Tangible Non-Residential Fixed Capital Stocks in National Prices and Standardized Estimates* (all figures are adjusted to a mid-year basis)

	Standardized estimate		Official estimates		Ratio official/ standardized	
	Gross	Net	Gross	Net	Gross	Net
Germany (billion 1980 DM)						
1950	640.8	326.1	935.2	563.5	146	172
1973	2,834.6	1,823.3	3,263.5	2,333.5	115	128
1987	4,537.8	2,637.9	5,132.7	3,514.1	113	133
Annual average compound growth rates						
1950–73	6.7	7.8	5.6	6.4		
1973–6	3.4	2.7	3.3	3.0		
UK (billion 1985 £)						
1950	185.4	99.1	341.6	190.2	184	192
1973	582.7	359.8	745.7	474.6	128	132
1987	877.8	494.4	1,055.2	642.9	120	130
Annual average compound growth rates						
1950–73	5.1	5.8	3.5	4.1		
1973–87	3.0	2.3	2.5	2.2		
USA (billion 1982 $)						
1950	2,902.2	1,550.3	2,843.9	1,538.1	98	99
1973	6,053.2	3,684.4	6,180.4	3,774.1	102	102
1987	9,535.5	5,286.1	9,332.9	5,358.0	98	101
Annual average compound growth rates						
1950–73	3.2	3.8	3.4	4.0		
1973–87	3.3	2.6	3.0	2.5		

Sources: See country notes below.

formation figures as those used to construct the official capital stock series. For the UK I used the same investment series as those in the official estimates for 1948–87 and the standard source (Feinstein) for pre-war years. For other countries I used national accounts for post-war years and the standard historical sources, so I presume that they closely replicate those officially used. However, there may be differences in the years chosen for weighting and deflation, or in the assumptions made about war damage. Any defects of this kind in the standardized (or the official) series would be remediable with more detailed public availability of the sources used in constructing the official series.

There are of course differences in the age at which assets are scrapped in different countries, so it could be argued that standardization is not warranted. However, the inter-country variation is quite unlikely to be as large as the different authorities assume, and more fundamentally, as our purpose is to measure assets as prices prevailing in the lead country, I would argue that the US price is zero for the old vintages of capital which the British and German statisticians include in their stocks and which US statisticians assume to have been scrapped. Ultimately, the case for a standardized approach is that it provides a much cleaner reflection of differences in historical patterns of accumulation.

SOURCES FOR STANDARDIZED ESTIMATES

For the USA, annual investment in structure since 1850 and in equipment since 1878 at 1982 prices was taken from *Fixed Reproducible Tangible Wealth in the United States 1925–85*, Bureau of Economic Analysis, US Dept. of Commerce, 1987, pp. 333–4 and 363–9, with updating supplied by John C. Musgrave. These are the same figures used to construct the official capital stock estimates.

For other countries, the detail publicly available on the investment flows used to construct official capital stock estimates is more meagre than in the USA. Annual investment since 1950 and investment deflators for France and the Netherlands were taken from OECD, *National Accounts, 1950–64*, from the 1950–68 edition at 1963 prices, linked to 1964–72 from the 1963–80 edition at 1975 prices, linked to recent years at 1980 prices from OECD, *National Accounts, 1975–87*. For years prior to 1950 the following sources were used:

France: 1938–50 was backcast from estimates in A. Maddison, 'Explaining Economic Growth', *Banca Nazionale del Lavoro Quarterly Review*, Sept. 1972, assuming both types of investment to move parallel. 1910–38 volume movement for structures and 1935–8 for equipment from J.-J. Carré, P. Dubois, and E. Malinvaud, *La Croissance française*, Seuil, Paris, 1972, p. 652, with interpolation for 1913–22. 8 per cent war damage was assumed for capital formation prior to and during each of the two world wars; see A. Maddison, 'Economic Policy and Performance in Europe 1913–70', *Fontana Economic History of Europe*, vol. v (2), Collins, London, 1976, p. 472. This means that pre-1919 investment was reduced by 16 per cent and 1919–45 capital formation by 8 per cent.

TABLE D.5. *Gross and Net Tangible Non-Residential Fixed Capital Stock, France, 1950–1987* (mid-year estimates, million $ at 1985 US relative prices)

	Gross stock of non-residential structures	Gross stock of machinery and equipment	Net stock of non-residential structures	Net stock of machinery and equipment
1950	236,038	54,972	115,882	30,825
1960	289,552	134,656	153,654	77,549
1973	532,917	406,715	367,374	239,998
1987	985,420	767,561	612,253	395,972

Note: Investment, retirements, and depreciation were measured in 1980 francs, converted to 1985 francs by a coefficient of 1.4623 for structures and 1.5465 for equipment. Conversion to 1985 dollars was made by the Eurostat PPPs, 0.13844 for structures and 0.15148 for equipment. (US cents per franc).

TABLE D.6. *Gross and Net Tangible Non-Residential Fixed Capital Stock, Germany, 1950–1987* (mid-year estimates, million $ at 1985 US relative prices)

	Gross stock of non-residential structures	Gross stock of machinery and equipment	Net stock of non-residential structures	Net stock of machinery and equipment
1950	267,367	77,411	139,130	36,722
1960	428,151	176,731	271,921	110,956
1973	980,639	519,384	678,962	291,871
1987	1,652,728	758,873	989,599	415,911

Note: Investment, retirements, and depreciation were measured in 1980 DM, converted to 1985 DM by a coefficient of 1.08727 for structures and 1.17866 for equipment. Conversion to 1985 dollars was made by the Eurostat PPPs, 0.51065 for structures and 0.41245 for equipment. (US cents per DM).

TABLE D.7. *Gross and Net Tangible Non-Residential Fixed Capital Stock, Japan, 1890–1987* (mid-year estimates, million $ at 1985 US relative prices)

	Gross stock of non-residential structures	Gross stock of machinery and equipment	Net stock of non-residential structures	Net stock of machinery and equipment
1890	21,856	7,666	14,524	4,258
1913	40,470	17,826	26,895	10,346
1929	81,069	47,726	54,410	23,216
1938	120,451	55,460	77,187	32,040
1950	137,208	98,637	90,373	47,015
1960	242,174	118,637	171,763	72,684
1973	1,099,011	641,776	898,111	386,685
1987	3,257,949	1,392,879	2,364,490	799,957

Note: Investment, retirements, and depreciation were measured in 1980 yen, converted to 1985 yen by a coefficient of 1.12845 for structures and 0.87742 for equipment. Conversion to 1985 dollars was made by the Eurostat PPPs, 0.004269 for structures and 0.004249 for equipment. (US cents per yen).

Germany: for 1960 onwards investment was supplied by the Federal Statistical Office, and for 1950–60 from OECD sources. Earlier years from W. Kirner, *Zeitreihen für das Anlagevermögen der Wirtschafts-bereiche in der Bundesrepublik Deutschland*, Duncker and Humblot, Berlin, 1968, pp. 103–5 for structures (back to 1850) and pp. 106–7 for equipment (partial from 1900, complete from 1930). Kirner's investment figures before 1946 are already adjusted to exclude war damage (about 16 per cent).

Japan: 1885–1940 investment by type of asset at 1934–6 prices from K. Ohkawa and M. Shinohara, *Patterns of Japanese Economic Development*, Yale, 1979, pp. 357–61. 1940–52 at 1934–6 prices (adjusted to a calendar year basis) from K. Ohkawa and H. Rosovsky, *Japanese Economic Growth*, Stanford, 1973, pp. 292–3. 1952–70 at 1965 prices from Ohkawa and Shinohara, p. 365. 1970–87 at 1980 prices from OECD, *National Accounts*, various issues. For 1941–69 the above sources do not show a breakdown by type of asset at constant prices. As a proxy for total machinery and equipment I used Ohkawa and Shinohara, p. 363, figures for the 1952–70 movement of government investment in machinery and equipment (current prices, adjusted to constant prices by their implicit deflator for total fixed investment). 1941–5 machinery and equipment investment was assumed to be the same proportion of total non-residential fixed investment as in 1940 and for 1946–51 it was assumed to be the same proportion of the total as in 1952. War damage was taken to be 25.7 per cent of pre-1946 investment, see *Hundred Year Statistics of the Japan Economy*, Bank of Japan, 1966, p. 27.

Netherlands: gross investment 1921–39 at 1938 prices from C. A. van Bochove and T. A. Huitker, 'Main National Accounting Series 1900–1986', *National Accounts: Occasional Paper, NA-017*, CBS, The Hague, 1987, with a breakdown by type of asset supplied by CBS. 1910–21 and 1939–50 figures were not available from CBS and as a proxy I assumed both types of asset to move parallel to machinery and equipment investment by enterprises at 1970 prices from Appendix 7.2 of H. den Hartog and T. S. Tjan, 'A Clay-Clay Vintage Model Approach for Sectors of Industry in the Netherlands', Central Planning Bureau, The Hague, Sept. 1979, mimeographed. This source puts war damage at 35 per cent, probably based on the Dutch official estimates made to back claims for reparations — *Memorandum van de Nederlandsche Regeering inzake de door Nederland van Duitschland te eischen schadevergoeding*, The Hague, 1945. However, this is way out of line with the estimate of 16 per cent war damage for Germany. I therefore assumed war damage of 10 per cent of pre-1946 investment.

UK: fixed asset formation at 1900 prices by type of asset for 1851–1920 from C. H. Feinstein and S. Pollard, *Studies in Capital Formation in the*

TABLE D.8. *Gross and Net Tangible Non-Residential Fixed Capital Stock, Netherlands, 1950–1987* (mid-year estimates, million $ at 1985 US relative prices)

	Gross stock of non-residential structures	Gross stock of machinery and equipment	Net stock of non-residential structures	Net stock of machinery and equipment
1950	70,031	13,114	35,617	7,554
1960	103,237	39,147	63,106	23,138
1973	208,112	98,104	144,239	55,712
1987	325,145	159,429	185,338	84,234

Note: Investment, retirements, and depreciation were measured in 1980 guilders, converted to 1985 guilders by a coefficient of 1.10626 for structures and 1.2153 for equipment. Conversion to 1985 dollars was made by the Eurostat PPPs, 0.41789 for structures and 0.36357 for equipment. (US cents per guilder).

TABLE D.9. *Gross and Net Tangible Non-Residential Fixed Capital Stock, UK, 1890–1987 (mid-year estimates, million $ at 1985 US relative prices adjusted for territorial change)*

	Gross stock of non-residential structures	Gross stock of machinery and equipment	Net stock of non-residential structures	Net stock of machinery and equipment
1890	88,910	25,803	48,348	13,026
1913	121,175	60,393	64,164	29,881
1929	131,019	99,760	66,741	56,709
1938	146,778	133,958	77,462	70,267
1950	147,567	164,301	80,105	86,205
1960	222,620	284,436	136,170	158,069
1973	462,754	517,710	316,198	279,130
1987	755,928	702,750	448,285	365,006

Note: Investment, retirements, and depreciation were measured in 1985 pounds. Conversion to 1985 dollars was made by the Eurostat PPPs, 1.45943 for structures and 1.9494 for equipment. (US cents per pound).

TABLE D.10. *Gross and Net Tangible Non-Residential Fixed Capital Stock, USA, 1890–1989* (mid-year estimates, million $ at 1985 US prices)

	Gross stock of non-residential structures	Gross stock of machinery and equipment	Net stock of non-residential structures	Net stock of machinery and equipment
1890	383,240	26,330	257,068	15,947
1913	1,151,871	225,708	729,664	123,076
1929	1,789,482	405,346	1,033,056	213,147
1938	2,014,294	371,156	1,065,015	174,875
1950	2,191,938	774,558	1,134,996	448,299
1960	2,853,971	1,131,493	1,624,054	585,549
1973	4,292,798	1,883,735	2,694,877	1,067,762
1987	6,191,949	3,515,751	3,485,491	1,898,114
1989	6,410,935	3,865,067	3,541,719	2,126,326

Note: Investment, retirements, and depreciation were measured in 1982 dollars, converted to 1985 dollars by a coefficient of 1.03287 for structures and 0.992989 for equipment.

United Kingdom 1750–1820, Oxford, 1988, pp. 446–7, linked to 1920–48 estimates at 1938 prices (adjusted for change in geographic boundaries of the UK) in C. H. Feinstein, *National Income, Expenditure and Output of the UK 1855–1965*, Cambridge, 1972, Table 40, linked to 1948–87 figures at 1985 prices supplied by the Central Statistical Office. War damage assumed to be 3 per cent of pre-1946 investment (see Maddison, 1976).

APPENDIX E

COST OF LIVING

For the period 1914–50, consumer price indices were generally derived from *Statistical Yearbooks* of the League of Nations and United Nations. For 1950–5, movements were from OECD, *General Statistics*, September 1962, pp. 59, 102, and 105 except for Australia, Finland, and Japan, which are from ILO, *Yearbook*, 1956. For 1955 onwards the figures are from OECD, *Main Economic Indicators* (the *Historical Statistics* volume and current issues). Otherwise the following sources were used:

Australia: 1870–1914 GDP deflator derived from N. G. Butlin, *Australian Domestic Product, Investment and Foreign Borrowing 1861–1938/9*, Cambridge, 1962, pp. 10, 11, 460–1.

Austria: 1874–1913 from D. F. Good, 'The Cost of Living in Austria: 1874–1913', *Journal of European Economic History*, Fall 1976; 1914–50 from B. R. Mitchell, *European Historical Statistics, 1750–1970*, Macmillan, London, 1975, pp. 743–5.

Belgium: 1870–1913 from J. Marczewski, 'Le Produit physique de l'économie française de 1789 à 1913', *Histoire quantitative de l'économie française*, INSEE, July 1965, pp. cxxv–cxxvi. 1914–19 from F. Baudhuin, *Histoire économique de la Belgique*, Bruylant, Brussels, 1944, pp. 37 and 97.

Canada: 1870–1914 GNP deflators from M. C. Urquhart, 'New Estimates of Gross National Product, Canada, 1870–1926: Some Implications for Canadian Development', pp. 30–1, in S. L. Engerman and R. E. Gallman, *Long-Term Factors in American Economic Growth*, NBER, New York, 1986. 1914–50 from M. C. Urquhart and K. A. H. Buckley, *Historical Statistics of Canada*, Macmillan, Toronto, 1965, pp. 303–4.

Denmark: 1870–1914 consumption deflator derived from K. Bjerke and N. Ussing, *Studier over Danmarks Nationalprodukt 1870–1950*, Gads, Copenhagen, 1958, pp. 148–9.

Finland: for 1870–1913, no consumer price index was available, so the wholesale price index was used, from H. Björkqvist, *Guldmyntfotens införande i Finland åren 1877–1878*, Bank of Finland, Helsinki, 1953, Table 16 (for 1870–7), and H. Björkqvist, *Prisrörelser och penningvärde i Finland under guldmyntfotsperioden 1878–1913*, Bank of Finland,

Helsinki, 1958, p. 259 (for 1878–1914); 1914–50 from B. R. Mitchell, *European Historical Statistics 1750–1970*, Macmillan, London, 1975, pp. 743–5.

France: 1820–1913 from J. Marczewski, 'Le produit physique', pp. civ, cxxv–cxxvi; 1913–14 link from *Annuaire statistique de la France 1954*, 2^e partie, p. 70; 1914–50 from *Annuaire statistique de la France 1960*, p. 387; 1914–20 (13 articles), 1921–50 (34 articles).

Germany: 1820–50 from B. R. Mitchell, *European Historical Statistics 1750–1970*, Macmillan, London, 1975; 1850–1913 private consumption deflator from W. C. Hoffmann and associates, *Das Wachstum der deutschen Wirtschaft seit der Mitte des 19. Jahrhunderts*, Springer, Berlin, 1965, pp. 599 and 601; 1913–23 from G. Bry, *Wages in Germany 1871–1945*, NBER, Princeton, 1960, pp. 440 and 445; 1924–50 from *Bevölkerung und Wirtschaft 1872–1972*, Statistisches Bundesamt, Wiesbaden, 1972, p. 250.

Italy: 1870–1950 from G. Fua, *Lo sviluppo economico in Italia*, Angeli, Milan, 1975, p. 434.

Japan: 1879–1944 from *Historical Statistics of Japanese Economy*, Bank of Japan, 1962, p. 49; 1944–46 Tokyo retail price index, 1946–50 consumer price index from *Hundred-Year Statistics of the Japanese Economy*, Bank of Japan, 1966, pp. 80–1.

Netherlands: 1824–1900 from J. L. van Zanden, *De industrialisatie in Amsterdam, 1828–1914*, Bergen, 1987, pp. 139–41. 1900–50 from *Zestig jaren statistiek in tijdreeksen 1899–1959*, CBS, Zeist, 1959, p. 125.

Norway: 1870–1914 consumer price index from *National Accounts 1865–1960*, CBS, Oslo, 1970, pp. 352–4.

Sweden: 1820–70 movement from L. Jörberg, *A History of Prices in Sweden 1732–1914*, vol. ii, CWK, Gleerup, 1972, pp. 185 and 350; 1870–1913 from G. Myrdal, *The Cost of Living in Sweden 1830–1930*, Stockholm, 1933.

Switzerland: 1892–1917 from E. Notz, *Die säkulare Entwicklung der Kaufkraft des Geldes*, Fischer, Jena, 1925, pp. 91–3 (31 articles).

UK: 1820–46 Rousseaux price index from B. R. Mitchell, *Abstract of British Historical Statistics*; 1846–1913, Bowley's index as cited by Marczewski, 'Le produit physique'; 1913–50 from *The British Economy: Key Statistics 1900–1970*, p. 8 (retail prices, all items).

USA: 1820–1950 consumer price index (BLS) from *Historical Statistics of the United States, Colonial times to 1970*, pp. 210–11.

TABLE E.1. *Consumer Price Indices, 1820–1860 (1913 = 100)*

	France	Germany	Netherlands	Sweden	UK	USA
1820	86.0	62.5	60.1[a]	55.9	125.8	141.4
1850	73.8	56.8	67.4		92.1	84.2
1860	86.4	70.3	102.1		110.8	90.9

[a] 1824.

TABLE E.2. *Consumer Price Indices, Annual Data, 1870–1914 (1913 = 100)*

	Australia	Austria	Belgium	Canada	Denmark	Finland	France	Germany
1870	89.4		101.0	77.6	110.9	78	94.1	76.5
1871	89.2		103.0	79.9	109.7	81	101.9	79.8
1872	96.1		106.8	87.3	108.9	87	99.0	85.0
1873	101.1		115.4	86.6	113.6	89	98.6	89.5
1874	98.8	97.4	109.7	84.3	116.2	101	97.9	90.3
1875	96.5	91.1	105.8	80.6	117.2	102	94.9	85.7
1876	95.9	93.1	111.5	80.6	117.0	105	98.2	86.1
1877	94.2	93.0	111.5	77.6	111.3	101	99.4	84.5
1878	90.1	89.0	106.8	76.1	106.2	84	96.3	82.4
1879	90.4	87.7	103.8	75.4	104.5	76	95.2	79.8
1880	89.4	88.0	99.1	77.6	108.6	89	98.4	83.7
1881	88.7	86.8	99.1	80.6	107.6	94	97.8	81.9
1882	97.7	87.6	98.1	84.3	104.0	87	96.3	82.8
1883	95.0	86.7	99.1	83.6	102.8	84	99.0	80.2
1884	92.6	84.5	92.3	73.9	98.9	83	97.1	79.2
1885	93.0	80.5	89.4	74.6	97.1	77	94.7	77.5
1886	90.3	79.0	81.8	74.6	93.7	72	93.3	77.0
1887	90.0	76.9	86.6	79.1	91.8	70	90.7	76.4
1888	92.0	76.8	84.6	76.1	91.1	70	91.7	77.1
1889	93.3	78.3	86.6	79.1	92.7	77	91.9	79.8
1890	93.7	78.5	90.4	77.6	95.1	77	91.1	82.3

Year								
1891	85.5	80.4	90.4	77.6	97.4	82	92.7	83.2
1892	82.3	77.8	86.6	77.6	96.1	82	91.9	82.8
1893	77.9	77.5	83.6	76.1	93.7	79	90.4	78.1
1894	73.0	77.1	81.8	69.4	90.9	73	93.3	77.0
1895	73.4	78.4	79.8	67.9	91.8	73	90.9	77.2
1896	77.4	75.3	76.9	70.1	88.9	74	89.4	78.7
1897	78.8	77.2	77.9	70.9	84.2	77	87.0	80.1
1898	79.2	79.7	77.9	73.1	89.2	80	88.2	80.2
1899	80.6	79.7	78.9	71.6	87.9	83	89.4	81.9
1900	78.5	78.8	89.4	74.6	93.0	87	89.4	84.4
1901	84.0	77.8	91.4	75.4	95.9	85	89.9	84.3
1902	84.6	76.8	89.4	77.6	95.1	85	88.8	84.5
1903	82.9	77.0	89.4	79.1	94.5	83	88.5	84.5
1904	82.9	79.3	77.9	79.9	92.1	83	87.2	85.0
1905	85.4	86.1	79.8	81.3	93.9	82	86.8	87.7
1906	87.3	86.8	87.5	82.8	95.5	87	88.2	92.5
1907	89.1	88.1	90.4	88.8	94.4	93	89.4	92.0
1908	88.2	91.5	91.4	89.6	96.2	91	91.5	91.5
1909	87.8	94.1	90.4	90.3	97.0	93	91.2	94.5
1910	89.2	93.8	92.3	91.0	97.5	95	101.6	98.4
1911	93.3	99.5	98.1	94.0	97.3	97	92.5	97.9
1912	97.0	100.5	104.8	95.5	97.9	100	101.3	99.3
1913	100.0	100.0	100.0	100.0	100.0	100	100.0	100.0
1914	108.6			99.3	104.1	103	99.6	103.0

TABLE E.2. (Cont.) *Consumer Price Indices, Annual Data, 1870–1914 (1913 = 100)*

	Italy	Japan	Netherlands	Norway	Sweden	Switzerland	UK	USA
1870	78.6		96.9	76.3	81.3		107.9	127.9
1871	80.1		102.0	77.2	83.4		110.8	121.2
1872	90.8		107.3	83.3	86.7		117.7	121.2
1873	99.5		111.6	89.8	93.6		119.6	121.2
1874	96.6		115.0	93.5	97.0		112.8	114.5
1875	80.6		107.3	90.2	96.3		108.8	111.1
1876	83.0		108.1	89.8	96.6		107.9	107.7
1877	92.2		112.5	89.8	96.1		107.9	107.7
1878	85.4		112.5	80.5	89.8		102.0	97.6
1879	84.0	38.5	108.1	75.8	84.2		99.0	94.3
1880	86.9	44.3	109.0	81.4	88.6		102.9	97.6
1881	83.5	48.2	109.9	82.3	90.8		101.0	97.6
1882	84.0	45.4	106.4	82.8	88.3		100.0	97.6
1883	79.6	40.5	103.8	81.4	87.9		100.0	94.3
1884	78.6	37.5	102.9	78.1	84.7		95.1	90.9
1885	80.6	39.6	94.3	73.5	80.7		89.2	90.9
1886	83.0	37.1	90.0	72.1	76.8		87.2	90.9
1887	78.6	37.1	90.0	70.2	74.0		86.3	90.9
1888	77.7	36.7	88.2	71.6	76.7		86.3	90.9
1889	86.4	39.3	94.3	74.0	80.1		87.2	90.9
1890	85.4	43.3	95.2	74.0	81.8		87.2	90.9

Year	(1)	(2)	(3)	(4)	(5)	(6)	(7)	(8)
1891	85.4	42.1	97.7	77.2	84.3		87.2	90.9
1892	81.1	42.8	102.0	75.8	82.9	85.7	88.2	90.9
1893	80.6	43.6	91.7	74.4	79.5	83.9	87.2	90.9
1894	79.6	45.3	89.1	72.6	75.5	83.4	83.3	87.5
1895	82.5	49.3	86.5	72.1	76.8	81.2	81.4	84.2
1896	79.6	53.0	83.0	72.6	76.2	80.2	81.4	84.2
1897	80.6	61.6	84.8	72.6	78.7	81.4	83.3	84.2
1898	85.0	66.2	84.8	77.7	82.5	84.5	86.3	84.2
1899	83.0	61.6	84.8	80.0	86.1	81.9	84.3	84.2
1900	83.5	69.5	86.5	83.3	87.0	78.9	89.2	84.2
1901	80.1	68.2	91.0	82.3	84.9	79.9	88.2	84.2
1902	78.6	71.1	88.8	80.5	85.6	80.4	88.2	84.2
1903	84.0	75.0	88.8	79.5	87.1	82.0	89.2	87.5
1904	81.6	77.0	91.0	80.0	86.0	82.0	90.2	90.9
1905	82.5	79.3	91.0	80.9	87.9	84.8	90.2	90.9
1906	86.9	82.5	91.0	83.3	89.7	87.2	91.2	90.9
1907	86.9	88.6	92.1	86.5	94.3	89.7	93.1	94.3
1908	85.9	85.4	95.5	87.4	95.7	90.9	91.2	90.9
1909	88.3	84.0	94.4	87.4	94.9	92.4	92.1	90.9
1910	94.2	84.7	96.6	89.3	94.8	96.3	94.1	94.3
1911	96.1	90.9	97.8	92.1	93.5	98.6	95.1	94.3
1912	99.0	96.7	98.9	96.7	99.7	102.2	98.0	97.6
1913	100.0	100.0	100.0	100.0	100.0	100.0	100.0	100.0
1914	98.5	90.0	100.0	102.3	102.0	101.8	100.0	101.3

TABLE E.3. *Consumer Price Indices, Annual Data, 1914–1950 (1914 = 100)*

	Australia	Austria	Belgium	Canada	Denmark	Finland	France	Germany
1914	100	100	100	100	100	100	100	100
1915	113	158	156	103	116	100	120	125
1916	117	337	328	111	136	133	135	165
1917	116	672	746	134	155	244	163	246
1918	120	1,163	1,434	151	182	633	213	304
1919	133	2,492	344	166	211	922	268	403
1920	158	5,115		198	261	889	371	990
1921	150	9,981	366	163	232	1,055	333	1,301
1922	141	263,938	340	153	200	1,033	315	14,602
1923	151	76[a]	399	153	206	1,033	344	[b]
1924	149	86	469	149	216	1,055	395	128[a]
1925	153	97	498	152	211	1,100	424	140
1926	158	103	604	155	184	1,078	560	141
1927	157	106	743	153	177	1,089	593	148
1928	158	108	761	153	175	1,122	584	152
1929	161	111	805	155	173	1,111	621	154
1930	152	111	834	154	165	1,022	618	148
1931	137	106	749	138	156	944	609	136

Year								
1932	130	108	673	125	155	933	546	121
1933	126	106	665	120	160	890	520	118
1934	128	106	639	121	167	889	491	121
1935	130	106	643	122	172	890	440	123
1936	133	106	677	124	175	890	480	124
1937	137	106	739	128	181	944	611	125
1938	140	104	755	130	183	978	706	126
1939	144	103	748	129	188	989	763	126
1940	150	106	858	134	234	1,189	909	130
1941	158	107		142	272	1,400	1,062	133
1942	170	108		148	282	1,655	1,238	137
1943	177	109		151	286	1,866	1,578	138
1944	177	109		151	289	1,978	2,013	141
1945	177	116		151	293	2,778	2,778	145
1946	179	147	2,462	157	291	4,422	4,553	158
1947	187	288	2,506	170	298	5,744	7,273	169
1948	203	469	2,876	196	307	7,733	11,529	195
1949	222	611	2,785	203	312	7,843	12,830	209
1950	244	763	2,754	209	331	8,948	13,731	196

[a] Linked to base year via gold price.
[b] Figure was 15,437,000 million.

TABLE E.3. (Cont.) *Consumer Price Indices, Annual Data, 1914–1950 (1914 = 100)*

	Italy	Japan	Netherlands	Norway	Sweden	Switzerland	UK	USA
1914	100	100	100	100	100	100	100	100
1915	109	94	115	117	115	115	124	102
1916	155	102	128	146	130	134	143	115
1917	224	122	136	190	159	171	176	138
1918	289	161	162	253	219	204	200	169
1919	331	213	176	275	257	222	219	193
1920	467	224	194	300	269	224	248	194
1921	467	210	169	277	247	200	224	169
1922	467	210	149	231	198	164	181	165
1923	476	206	144	218	178	164	176	168
1924	481	207	145	239	174	169	176	168
1925	580	211	144	243	177	168	176	173
1926	618	203	138	206	173	162	171	171
1927	547	202	138	186	171	160	167	167
1928	511	196	139	173	172	161	167	165
1929	503	192	138	166	170	161	167	165
1930	476	174	133	161	165	158	157	161

Year								
1931	421	159	125	153	160	150	148	147
1932	394	160	116	150	156	138	143	131
1933	363	167	115	147	155	131	143	124
1934	370	171	115	148	155	129	143	129
1935	381	174	111	151	156	128	143	132
1936	409	178	106	155	158	130	148	134
1937	454	193	112	166	163	137	152	138
1938	493	214	114	172	167	137	157	136
1939	516	231	115	174	172	138	162	134
1940	636	295	132	203	194	151	181	135
1941	788	347	151	238	219	174	200	142
1942	1,092	458	162	252	238	193	214	157
1943	1,753	544	167	258	240	203	224	167
1944	4,292	688	172	261	243	208	224	169
1945	9,340	1,011	198	266	243	210	233	174
1946	12,739	60,048	216	272	243	208	243	187
1947	22,020	129,104	224	274	253	217	257	216
1948	23,616	235,896	232	272	264	224	271	231
1949	23,665	311,386	246	274	267	222	281	229
1950	24,404	289,944	270	285	271	219	291	231

TABLE E.4. *Consumer Price Indices, Annual Data, 1950–1989 (1950 = 100.0)*

	Australia	Austria	Belgium	Canada	Denmark	Finland	France	Germany
1950	100.0	100.0	100.0	100.0	100.0	100.0	100.0	100.0
1951	120.6	127.7	108.9	110.4	112.5	120.5	116.9	107.8
1952	141.2	149.5	110.3	113.2	114.8	125.6	130.7	110.1
1953	147.1	141.4	110.0	112.2	113.6	128.2	129.2	108.1
1954	148.5	146.7	111.2	112.9	114.8	128.2	128.8	108.3
1955	152.9	147.8	110.9	113.1	122.7	124.4	130.2	110.1
1956	162.4	152.5	114.0	114.7	128.9	138.1	132.8	113.0
1957	166.5	156.0	117.5	118.4	129.4	154.2	137.4	115.2
1958	168.8	159.5	119.2	121.4	130.1	163.2	158.1	117.8
1959	172.0	161.1	120.6	122.8	132.5	166.5	167.9	118.9
1960	178.4	164.1	121.0	124.1	134.2	171.3	173.8	120.5
1961	182.9	170.0	122.1	124.7	139.6	174.3	179.5	123.3
1962	182.2	177.5	123.9	126.2	150.1	182.4	188.3	127.1
1963	183.4	182.4	126.5	128.5	158.2	191.9	197.3	130.9
1964	187.7	189.4	131.9	130.7	163.9	210.9	204.1	133.8
1965	195.2	198.7	137.1	134.0	174.4	222.0	209.2	138.4
1966	200.9	203.1	142.9	139.0	186.1	229.8	214.9	143.3
1967	207.3	211.1	147.0	143.8	199.0	242.7	220.6	145.3

1968	212.9	217.2	151.1	149.8	216.1	265.8	230.7	149.1
1969	218.9	223.7	156.6	156.6	225.2	271.4	245.4	151.9
1970	227.5	233.5	162.8	161.8	238.3	278.9	258.3	157.1
1971	241.4	244.5	169.8	166.5	252.1	297.1	272.5	165.4
1972	255.5	259.9	179.1	174.4	268.8	318.0	288.5	174.5
1973	279.6	279.5	191.6	187.7	293.8	351.4	309.7	186.6
1974	321.9	306.1	215.9	208.1	337.9	412.8	352.1	199.6
1975	370.4	332.0	243.4	230.6	375.1	485.3	393.1	211.6
1976	420.4	356.2	265.8	247.9	408.1	553.2	430.8	221.1
1977	472.3	375.8	284.5	267.7	453.9	626.0	471.3	229.8
1978	509.7	389.4	297.2	291.7	498.9	674.6	514.2	235.7
1979	553.0	405.9	311.3	316.5	547.5	725.2	569.5	246.3
1980	609.4	431.9	331.8	348.8	614.8	809.3	647.0	259.8
1981	667.9	461.2	357.1	392.4	686.8	906.4	733.6	276.2
1982	741.4	486.2	388.1	434.8	756.2	887.0	820.2	290.9
1983	816.3	502.2	418.0	460.0	808.3	960.6	899.0	300.5
1984	848.1	530.3	444.4	479.8	859.3	1,028.8	965.5	307.7
1985	905.8	547.3	466.1	498.9	899.6	1,089.5	1,021.5	314.5
1986	987.3	556.6	472.2	519.4	932.0	1,121.1	1,049.1	313.8
1987	1,071.6	564.3	479.6	542.4	969.8	1,166.9	1,081.7	313.8
1988	1,148.8	575.6	485.4	564.1	1,014.4	1,226.4	1,110.9	317.9
1989	1,236.1	590.0	500.4	592.3	1,063.1	1,307.4	1,150.9	326.8

TABLE E.4. (Cont.) Consumer Price Indices, Annual Data, 1950–1989 (1950 = 100.0)

	Italy	Japan	Netherlands	Norway	Sweden	Switzerland	UK	USA
1950	100.0	100.0	100.0	100.0	100.0	100.0	100.0	100.0
1951	109.3	115.6	109.0	116.9	115.2	104.9	108.8	107.9
1952	114.0	122.1	109.0	127.3	124.1	107.6	119.2	110.3
1953	116.3	129.9	109.0	129.9	126.6	106.8	122.9	111.2
1954	119.8	137.7	113.3	135.1	127.8	107.6	125.1	111.6
1955	123.3	136.4	115.5	136.4	131.6	108.5	130.7	111.3
1956	127.4	136.9	117.7	141.5	138.2	110.1	137.1	112.9
1957	129.2	141.1	125.3	145.4	144.0	112.3	142.2	116.9
1958	132.8	140.5	127.5	152.4	150.4	114.2	146.6	120.2
1959	132.2	141.8	128.5	155.8	151.6	113.6	147.3	121.0
1960	135.2	147.0	134.9	156.3	157.9	115.2	148.8	122.9
1961	137.9	154.8	136.3	160.4	161.2	117.3	153.9	124.2
1962	144.5	165.4	138.9	168.7	168.9	122.4	160.6	125.7
1963	155.2	179.4	144.3	173.0	173.8	126.6	163.7	127.3
1964	164.5	186.4	152.0	183.0	179.7	130.6	169.0	128.9
1965	171.8	198.9	158.2	190.8	188.6	135.0	177.2	131.1
1966	175.8	209.0	167.2	197.1	200.8	141.4	184.1	134.8
1967	181.6	217.3	173.0	205.8	209.5	147.0	188.7	138.7

1968	183.9	229.0	179.4	212.8	213.5	150.6	197.6	144.5
1969	188.7	240.9	192.8	219.4	219.1	154.4	208.2	152.3
1970	198.2	259.3	199.8	242.7	234.6	160.0	221.5	161.3
1971	207.7	275.1	214.8	257.7	252.0	170.6	242.3	168.2
1972	219.6	287.6	231.6	276.4	267.0	181.9	259.6	173.7
1973	243.4	321.3	250.1	296.1	285.0	197.8	283.5	184.5
1974	290.0	399.9	274.3	325.2	315.6	217.2	328.7	204.8
1975	339.1	447.0	302.3	364.1	344.9	231.7	408.4	223.6
1976	396.0	489.0	328.9	396.9	380.1	235.6	476.0	236.6
1977	468.9	528.8	350.1	433.3	424.2	238.7	551.3	251.0
1978	525.9	551.2	364.3	469.7	465.6	241.8	597.1	271.0
1979	603.5	569.4	379.8	492.0	499.1	249.6	670.2	295.4
1980	731.3	615.0	404.5	545.6	567.5	259.6	790.8	335.3
1981	861.5	645.1	431.6	620.3	636.2	276.5	884.9	369.8
1982	1,003.6	662.5	457.5	690.4	690.9	291.9	961.0	392.4
1983	1,150.1	675.1	470.3	748.4	752.4	300.7	1,005.2	404.9
1984	1,274.4	690.0	485.9	794.8	812.6	310.0	1,055.5	422.3
1985	1,391.6	704.5	497.0	840.2	873.0	320.5	1,119.9	437.1
1986	1,476.5	708.7	500.0	900.6	910.5	322.8	1,157.9	445.4
1987	1,544.7	709.4	494.0	978.8	948.1	327.6	1,206.1	461.6
1988	1,621.9	714.4	497.5	1,044.4	1,003.1	333.8	1,265.2	480.5
1989	1,729.0	730.8	502.9	1,092.4	1,067.3	344.5	1,363.9	503.6

APPENDIX F

THE VOLUME OF MERCHANDISE EXPORTS

Unless otherwise specified, the estimates for 1870–1950 are from A. Maddison, 'Growth and Fluctuation in the World Economy', *Banca Nazionale del Lavoro Quarterly Review*, June 1962, and for 1950–1973 (and in some cases 1946–9) from the UN *Yearbooks of International Trade Statistics* and the UN *Monthly Bulletins of Statistics*. For 1973 onwards the indices are from OECD sources.

The figures are on a calendar year basis. Up to 1950 the indices refer to special exports (excluding re-exports of merchandise). They exclude exports of services unless otherwise specified.

For the following countries and periods, the sources used are listed below.

Australia: 1871–1913 export values from N. G. Butlin, *Australian Domestic Product, Investment and Foreign Borrowing 1861–1938/9*, Cambridge, 1962, pp. 410–11, 436, 438, and 441 (excluding gold); export deflator from W. Vamplew, *Australians: Historical Statistics*, Fairfax, Syme, and Weldon, Broadway, 1987, p. 190, and C. P. Kindleberger, *The Terms of Trade*, Chapman & Hall, London, 1956, p. 157 (Wilson's index). 1913–36 exports at constant prices from M. W. Butlin, *A Preliminary Annual Database*, Reserve Bank, 1977; 1936–50 UN *Yearbook*. 1913 onwards adjusted to a calendar year basis.

Austria: 1870–1913 from P. Bairoch, *Commerce extérieur et développement économique de l'Europe au XIXᵉ siècle*, Mouton, Paris, 1976, p. 76. 1913–37 from A. Kausel, *Österreichs Volkseinkommen 1913 bis 1963*, p. 4 (includes exports of services); 1937–50 from UN, *Yearbook of International Trade Statistics*, 1959. The Austrian figures refer throughout to the present territory, whose 1913 exports were 47.77 per cent of those of actual 1913 Austria (see Kausel, *Österreichs Volkseinkommen*, p. 25, and Maddison, 'Growth and Fluctuation', p. 56).

Belgium: 1831–70 from S. Capelle, 'Le volume de commerce extérieur de la Belgique 1830–1913', *Bulletin de l'Institut de Recherches Économiques*, Louvain, 1938, p. 54; 1870–90 values from B. R. Mitchell, *European Historical Statistics 1750–1970*, Macmillan, London, 1975, p. 489, converted into dollars at 1913 exchange rates and deflated by dollar unit value index in Maddison, 'Growth and Fluctuation'; 1890–1950 volume from Maddison, 'Growth and Fluctuation'; 1939–45 roughly estimated from UN, *Yearbook of International Trade Statistics*, 1959, by deflating dollar value of exports by US export price index and adjusting to fit within the 1938–46 volume movement. From May 1922 to August 1940 and since May 1945 the statistics refer to the Belgium–Luxembourg customs area.

Canada: 1913–50 from M. C. Urquhart and K. A. H. Buckley, *Historical Statistics of Canada*, Cambridge, 1965, p. 178 (for 1913–25 adjusted to a calendar year basis); pp. 183–4 for the 1913–26 link and p. 179 for 1926–50.

Denmark: 1844–74 export values from S. A. Hansen, *Økonomisk vækst i Danmark*, vol. ii, Akademisk Forlag, Copenhagen, 1974, pp. 254–7, divided by export price deflator, pp. 293–4. Hansen gives figures starting in 1818 but they are incomplete before 1844. Export values 1875–1950 from K. Bjerke and N. Ussing, *Studier over Danmarks Nationalprodukt 1870–1950*, Gads, Copenhagen, 1958, pp. 152–3, divided by the export price index for those years in A. Olgaard, *Growth Productivity and Relative Prices*, Copenhagen, 1966, p. 242.

Finland: 1870–1949 from E. Pihkala, *Finland's Foreign Trade 1860–1917*, Bank of Finland, Helsinki, 1969, p. 63, and H. Oksanen and E. Pihkala, *Finland's Foreign Trade 1917–1949*, Bank of Finland, Helsinki, 1975, p. 39, *Studies in Finland's Economic Growth*, Bank of Finland, Helsinki, 1975.

France: 1715–87 from E. Levasseur, *Histoire du commerce de la France, première partie avant 1789*, Rousseau, Paris, 1911, p. 518, deflated by 60 per cent for price changes (see F. Crouzet, 'England and France in the Eighteenth Century: A Comparative Analysis of Two Economic Growths', in R. M. Hartwell (ed.), *The Causes of the Industrial Revolution in England*, Methuen, London, 1967, p. 146); 1800–10 from a chart in Levasseur, *Histoire du commerce de la France*, vol. ii (1912); it was assumed that exports in 1800 were at the same level as in 1787; 1810–70 from M. Lévy-Leboyer, 'La croissance économique en France au XIXe siècle, *Annales*, July–Aug. 1968; 1913–50 from *Annuaire statistique de la France 1966, résumé rétrospectif*, pp. 350, 360–1. The figures for 1913–18 were derived from a volume index of exports and imports combined, based on share of exports in total trade value.

Germany: 1836–1913 and 1926–7 from W. G. Hoffmann and associates, *Das Wachstum der deutschen Wirtschaft seit der Mitte des 19. Jahrhunderts*, Springer, Berlin, 1965, pp. 530–1. For 1914–18 the figures are estimates based on the assumption that exports in the first eight months of 1914 were the same as in 1913, and pro-rating Germany's total wartime exports in what seemed a plausible pattern. The figures on total wartime exports

(12 billion gold marks) are from C. Bresciani-Turroni. *The Economics of Inflation*, Allen & Unwin, London, 1968, p. 85; 1919 was estimated relative to 1920 from information given by Bresciani-Turroni, pp. 229 and 248; 1920–5 volume movement from *Statistik des Deutschen Reiches*, vol. 329 *I-Der auswärtige Handel Deutschlands im Jahre 1925*, p. 5; 1928–43 from *Statistisches Handbuch von Deutschland 1928–1944*, Länderrat des Amerikanischen Besatzungsgebiets, Munich, 1949, p. 395; 1944–50 derived from UN, *Yearbooks of International Trade Statistics*, 1954 and 1959 edns.; 1945–8 figures were assessed in dollars and deflated by the US export price index. It should be noted that, prior to 1906, the free ports of Hamburg, Cuxhaven, Bremerhaven, and Geestemünde were not included in the German customs area.

Italy: 1870–1950 merchandise export receipts divided by export price index in G. Fua, *Lo sviluppo economico in Italia*, Angeli, Milan, 1975, vol. iii, pp. 465–6 and 434–5.

Japan: 1870–3 from M. Baba and M. Tatemoto, 'Foreign Trade and Economic Growth in Japan: 1858–1937', in L. Klein and K. Ohkawa, *Economic Growth: The Japanese Experience since the Meiji Era*, Irwin, Homewood, Illinois, 1968, p. 167; 1873–1905 from K. Kojima, 'Japan's Foreign Trade and Economic Growth', *Annals of the Hitotsubashi Academy*, April 1958, pp. 166–7, whose figures exclude trade with colonies. Trade with colonies included by adjusting Kojima with trade ratios derived from *Historical Statistics of Japanese Economy*, Bank of Japan, 1962, pp. 89–90; 1905–50 from K. Ohkawa and H. Rosovsky, *Japanese Economic Growth*, Stanford, 1973, p. 302 (includes trade with colonies); 1945 derived from 1944–5 value movement shown by UN *Yearbook of International Trade Statistics*, divided by 1944–5 change in cost of living index.

Netherlands: 1872–1913 volume (excluding transit trade) from J. T. Lindblad and J. L. van Zanden, 'De buitenlandse handel van Nederland, 1872–1913', *Economisch en Sociaal-Historisch Jaarboek*, 1989, p. 262. 1913–21 value figures (from Mitchell, *European Historical Statistics*) deflated by the national income deflator. 1921–50 export volume from *Zeventig jaren statistiek in tijdreeksen*, The Hague, 1970, p. 92; 1940–5 roughly estimated from UN, *Yearbook of International Trade Statistics*, 1959, deflating dollar value of exports by the US export price index.

Norway: 1870–1939 and 1946–50 from *National Accounts 1865–1960*, Central Bureau of Statistics, Oslo, 1970, pp. 356–9 (includes exports of services); 1940–5 from UN, *Yearbook of International Trade Statistics*, 1959.

Sweden: 1870–1950 from O. Johansson, *The Gross Domestic Product of Sweden and its Composition 1861–1955*, Stockholm, 1967, pp. 140–1.

TABLE F.1. *Volume of Exports, 1720–1860* (1913 = 100)

	Austria	Belgium	Denmark	France	Germany	Italy	Switzerland	UK	USA
1720				1.536[f]				0.386	
1790				3.45[g]				1.19	0.64
1820	2.32			4.31		7.33	2.57	2.86	1.31
1830		2.19[a]		4.24				4.47	2.13
1840		4.00[b]	12.92[e]	6.51	4.3			7.38	4.08
1850		6.02[c]	13.59	10.56	5.9			13.07	4.07
1860		12.20[d]	11.60	19.55	9.8			22.68	9.61

[a] 1831. [b] 1843. [c] 1851. [d] 1861. [e] 1844. [f] 1715. [g] 1787.

Sources: As described in country notes, except for Austria, Italy, and Switzerland 1820–70, which are derived from value figures of M. G. Mulhall, *The Dictionary of Statistics*, Routledge, London, 1899, p. 128, deflated by the Rousseaux wholesale price index for the UK (see B. R. Mitchell, *Abstract of British Historical Statistics*, Cambridge, 1962, pp. 471–2).

TABLE F.2. *Volume of Exports, Annual Data, 1870–1913 (1913 = 100)*

	Australia	Austria	Belgium	Canada	Denmark	Finland	France	Germany
1870	13.4	23.1	16.9	17.9	21.0	19.4	31.1	17.7
1871	16.5		21.0	18.4	19.8	22.4	31.6	20.3
1872	21.6		24.0	20.5	23.5	n.a.	40.5	18.7
1873	17.0		25.8	21.6	21.6	n.a.	42.3	17.7
1874	21.3		25.4	19.8	23.6	27.4	43.4	19.0
1875	20.9		25.8	18.6	25.3	24.6	47.3	20.8
1876	23.6		25.3	19.0	27.3	27.9	43.7	21.2
1877	23.5		26.3	19.2	29.2	29.9	42.3	23.9
1878	24.5		27.5	19.0	21.9	25.6	41.6	26.1
1879	23.3		30.0	20.6	32.7	28.4	41.2	24.9
1880	31.3		31.2	22.9	35.0	36.8	43.5	22.4
1881	25.4		33.7	24.4	31.6	30.6	45.0	23.0
1882	28.3		35.0	23.6	29.3	35.3	45.6	23.9
1883	32.0		36.1	22.0	33.0	32.8	46.5	25.0
1884	30.7		37.9	21.8	31.2	34.1	46.1	26.4
1885	29.4		35.2	22.4	29.2	33.6	45.8	25.9
1886	25.7		37.4	23.1	32.3	29.1	48.7	28.1
1887	31.0		38.9	23.0	35.2	29.1	50.7	29.4
1888	34.8		38.9	22.4	34.4	32.6	48.2	29.6
1889	35.2		45.2	22.8	36.5	35.1	53.8	28.3
1890	35.1		45.0	23.6	42.2	34.1	53.0	29.8

Year							
1891	56.9	48.1	25.5	44.6	37.3	54.6	29.7
1892	46.0	44.4	28.1	42.7	33.6	56.5	29.4
1893	59.0	44.5	29.2	42.8	40.8	50.0	30.9
1894	60.6	41.8	29.6	51.5	47.3	53.9	31.7
1895	63.6	43.3	31.7	52.7	48.8	58.4	35.8
1896	61.5	44.9	35.7	53.1	52.2	59.6	37.1
1897	55.7	48.6	40.3	51.4	53.0	61.5	38.4
1898	58.0	51.7	41.2	52.2	52.2	56.7	39.6
1899	60.3	54.7	44.0	58.4	53.5	63.5	42.2
1900	58.2	53.9	48.0	56.4	52.0	61.6	44.7
1901	59.4	51.8	50.8	56.3	51.7	65.5	45.2
1902	45.8	54.6	54.3	62.1	55.0	68.7	48.9
1903	43.1	59.2	53.5	69.6	57.5	66.5	52.3
1904	66.0	60.0	51.2	73.7	61.9	68.1	53.6
1905	63.5	62.8	55.0	74.5	67.4	72.2	58.2
1906	70.8	73.7	58.0	71.7	73.4	75.0	64.9
1907	81.3	74.4	56.3	78.6	68.2	80.5	66.3
1908	73.2	66.8	55.4	83.9	65.9	77.3	65.8
1909	82.2	76.4	61.0	80.8	69.4	82.3	68.7
1910	97.9	92.6	62.3	84.7	75.1	88.0	77.4
1911	100.2	97.3	66.2	93.5	81.3	84.9	83.5
1912	91.1	106.4	78.3	97.6	87.3	93.8	89.7
1913	100.0	100.0	100.0	100.0	100.0	100.0	100.0

Note: Figures refer to the customs territory of the year specified.

TABLE F.2. (Cont.) *Volume of Exports, Annual Data, 1870–1913* (1913 = 100)

	Italy	Japan	Netherlands	Norway	Sweden	Switzerland	UK	USA
1870	38.7	3.0		26.1	26.7	(19.3)	31.1	13.0
1871	53.8	4.0		26.0	25.3		34.9	14.8
1872	52.2	4.0	39.9	31.0	26.7		36.2	15.0
1873	46.6	4.7	42.4	30.5	26.4		34.8	17.0
1874	41.1	4.8	40.2	30.3	26.0		34.6	17.8
1875	49.7	4.8	42.0	29.1	26.7		34.4	17.1
1876	59.5	6.6	41.7	31.6	30.3		33.5	19.6
1877	41.3	7.0	42.8	31.0	29.8		34.6	23.8
1878	47.2	7.7	42.1	29.8	29.4		34.8	28.7
1879	52.2	7.4	42.5	30.8	33.3		36.7	33.1
1880	51.7	6.9	40.5	34.9	34.4		41.2	35.0
1881	56.3	7.7	41.8	34.6	33.4		45.1	31.8
1882	54.7	10.0	43.5	35.1	36.5		45.6	28.4
1883	54.0	10.5	40.9	34.1	37.2		46.9	31.1
1884	55.1	9.4	43.9	36.1	37.9		47.3	30.5
1885	43.5	10.0	47.0	36.0	39.3		45.0	30.0
1886	49.4	12.4	46.9	36.9	37.7		46.9	33.1
1887	51.9	12.9	49.6	38.0	41.9		49.1	33.4
1888	46.1	17.8	49.1	41.3	45.1		52.2	30.7
1889	49.4	17.3	48.1	44.7	43.8		54.3	38.4
1890	43.2	13.2	52.5	45.6	45.4		55.1	40.4

1891	40.2	20.3	52.8	44.7	48.0		52.1	44.2
1892	47.1	20.3	51.0	45.8	46.8		50.2	45.8
1893	48.7	18.2	51.7	47.5	50.7		48.2	43.3
1894	55.1	22.2	50.0	45.0	50.6		50.3	46.5
1895	55.2	24.0	51.5	44.6	53.3		54.7	45.7
1896	55.0	21.3	58.3	47.1	57.1		57.7	56.4
1897	56.1	28.8	59.3	52.1	55.2		56.9	63.4
1898	59.1	27.2	60.5	48.3	51.4		56.5	73.2
1899	69.7	34.0	66.2	48.1	52.0		61.1	70.3
1900	64.4	29.3	64.7	49.4	54.0	63.3	58.6	72.8
1901	68.8	40.3	63.6	51.4	50.8		59.2	74.0
1902	74.2	40.9	64.0	56.1	56.2		62.8	66.9
1903	74.0	45.0	68.4	56.2	62.7		64.5	68.8
1904	77.7	48.4	71.4	59.1	59.9		65.9	67.0
1905	84.2	44.5	66.8	61.9	65.4		72.4	78.0
1906	89.4	57.6	71.3	66.8	69.5		77.8	80.5
1907	89.1	58.2	79.1	66.2	70.3		84.1	81.3
1908	79.2	56.7	80.5	68.9	67.6		77.4	78.4
1909	85.4	62.7	81.7	73.4	63.2		80.7	73.7
1910	91.7	74.3	85.1	78.2	78.0		88.0	73.1
1911	92.2	77.2	87.9	83.6	87.9		91.2	90.1
1912	95.0	89.4	98.5	89.5	96.3		96.3	101.1
1913	100.0	100.0	100.0	100.0	100.0	100.0	100.0	100.0

Note: Figures refer to the customs territory of the year specified.

TABLE F.3. *Volume of Exports, Annual Data, 1913–1950* (1913 = 100)

	Australia	Austria	Belgium	Canada	Denmark	Finland	France	Germany
1913	100.0	100.0	100.0	100.0	100.0	100.0	100.0	100.0
1914	91.2			100.3	123.7	68.2	71.3	(80.0)
1915	80.7			139.6	116.1	47.3	43.0	(50.0)
1916	86.1			188.5	117.8	48.8	45.0	(20.0)
1917	84.5			213.5	88.8	29.1	36.0	(20.0)
1918	79.9			154.8	39.9	11.4	28.0	(15.0)
1919	92.4			139.1	38.6	51.2	45.0	17.0
1920	93.2			118.5	77.9	77.6	86.0	36.7
1921	101.5			110.5	87.2	72.4	83.0	44.4
1922	106.8			143.0	96.6	102.7	86.0	61.3
1923	88.3			175.4	123.2	108.7	103.0	52.9
1924	85.8	75.6		174.1	142.5	124.9	119.0	50.8
1925	98.5	82.1	72.8	192.5	138.4	138.9	124.0	65.3
1926	104.5	76.7	76.4	194.0	147.1	142.4	134.0	72.4
1927	101.0	87.1	96.5	193.4	170.4	160.1	146.0	73.1
1928	105.5	91.7	107.8	221.1	179.3	155.7	148.0	82.5
1929	107.2	86.3	107.2	193.2	181.0	161.4	147.0	91.8
1930	113.7	77.3	92.8	172.7	200.0	140.9	132.0	87.0

Year								
1931	131.8	65.3	94.4	141.6	214.6	136.1	112.0	79.1
1932	141.3	48.4	73.5	131.3	217.4	142.2	86.0	54.6
1933	137.6	47.0	76.0	143.3	196.0	168.2	88.0	51.2
1934	139.0	48.2	78.5	164.6	182.8	186.0	91.0	45.8
1935	142.2	50.5	87.5	180.4	171.9	196.2	82.0	49.3
1936	139.3	52.2	97.9	221.1	175.5	218.0	78.0	54.4
1937	142.8	65.5	111.3	201.7	194.3	233.7	84.0	62.9
1938	153.8		102.4	192.0	184.8	201.5	91.0	57.0
1939	162.7		(103.0)	221.4	187.2	200.7	88.0	59.5
1940	153.1		(46.0)	255.2	158.8	62.3	40.0	38.5
1941	151.7		(19.0)	337.3	97.5	84.4	30.0	45.5
1942	140.7		(13.0)	464.2	78.4	85.9	45.0	46.3
1943	126.3		(20.0)	526.1	93.2	98.0	41.9	46.9
1944	131.1		(10.0)	549.6	95.1	68.0	25.5	31.1
1945	135.2		(5.0)	505.5	64.6	34.3	9.1	1.6
1946	150.3		39.1	312.6	120.5	94.9	36.4	4.1
1947	152.4	19.3	75.4	327.2	146.4	132.4	63.7	4.9
1948	156.5	35.7	92.9	332.2	138.4	145.9	74.6	9.4
1949	165.3	44.4	102.4	312.9	179.8	174.6	107.4	16.4
1950	158.7	66.6	111.8	311.0	239.5	199.5	149.2	34.8

Note: Figures refer to the customs territory of the year specified.

Table F.3. (Cont.) *Volume of Exports, Annual Data, 1913–1950* (1913 = 100)

	Italy	Japan	Netherlands	Norway[a]	Sweden	Switzerland	UK	USA
1913	100.0	100.0	100.0	100.0	100.0	100.0	100.0	100.0
1914	88.3	98.1	(81.3)	97.8	89.4		80.4	86.6
1915	89.2	130.1	(50.5)	103.7	137.8		56.8	135.7
1916	78.9	164.5	(35.0)	104.5	134.3		57.4	163.3
1917	42.5	194.6	(60.0)	70.4	76.1		44.6	142.2
1918	48.9	211.5	(23.8)	61.1	65.1		37.7	119.7
1919	75.1	168.3	(81.3)	67.1	66.5		54.7	146.6
1920	105.5	145.0	(87.4)	90.5	75.8		70.7	141.8
1921	70.7	128.3	83.0	76.0	59.1	62.6	49.3	113.4
1922	80.8	137.6	93.4	97.8	81.2	70.1	68.0	106.8
1923	93.7	117.6	103.7	104.5	79.7	74.4	74.7	108.2
1924	117.3	153.4	124.6	110.8	95.8	87.3	76.0	121.0
1925	127.4	183.6	134.9	121.8	106.0	89.7	74.7	128.0
1926	122.6	196.2	140.0	130.2	114.3	86.6	66.7	137.3
1927	116.1	218.5	160.8	140.8	136.3	98.2	77.3	147.8
1928	117.9	234.6	166.0	143.3	131.3	101.1	80.0	153.6
1929	122.7	257.9	171.2	167.1	156.1	100.7	81.3	158.2
1930	104.7	256.8	160.8	181.1	144.1	90.1	66.7	130.3
1931	98.3	265.8	150.4	159.4	117.4	77.1	50.7	105.9

Year								
1932	71.1	261.5	124.6	171.4	98.3	50.3	50.7	81.4
1933	65.3	346.0	114.1	180.6	110.6	52.6	52.0	82.6
1934	56.1	447.3	119.3	187.6	129.2	62.7	54.7	88.4
1935	49.3	525.0	119.3	194.1	131.7	62.4	60.0	93.1
1936	39.4	536.2	124.6	212.6	150.6	63.4	60.0	97.7
1937	73.9	642.5	150.4	233.2	158.1	70.1	65.3	125.7
1938	69.9	588.3	140.0	233.2	159.7	79.3	57.3	125.7
1939	81.7	526.5	129.7	243.0	176.1	77.8	53.3	131.5
1940	63.3	506.9	(78.8)	143.7	119.0	72.0	41.3	153.6
1941	75.8	549.5	(69.0)	104.5	109.8	70.5	28.0	182.7
1942	68.1	387.5	(53.5)	73.2	100.6	58.7	21.3	239.7
1943	67.4	345.4	(48.7)	73.2	93.9	49.9	16.0	350.3
1944	13.6	255.4	(36.7)	65.3	69.9	32.3	17.3	337.5
1945	4.4	30.9	(5.5)	41.8	108.4	45.5	26.7	229.3
1946	39.9	13.5	26.0	132.9	174.2	81.4	57.3	239.7
1947	73.2	32.7	57.0	179.5	200.6	93.3	62.7	320.0
1948	111.6	47.9	83.0	207.2	221.6	101.3	80.0	249.0
1949	115.7	98.4	124.6	225.1	225.6	101.3	88.0	254.9
1950	126.5	210.1	171.2	269.5	275.9	113.2	100.0	224.6

Note: Figures refer to the customs territory of the year specified.

[a] Includes exports of services.

Table F.4. *Volume of Exports, Annual Data, 1950–1989* (1913 = 100)

	Australia	Austria	Belgium	Canada	Denmark	Finland	France	Germany
1950	158.7	66.6	111.8	311.0	239.5	199.5	149.2	34.8
1951	146.7	77.8	130.5	345.6	276.3	249.4	169.6	50.3
1952	158.7	72.1	118.0	380.1	257.9	216.1	149.2	54.1
1953	177.9	100.0	130.5	368.6	285.5	249.4	156.0	61.9
1954	177.5	122.2	142.8	357.1	313.2	282.6	176.4	73.4
1955	182.7	133.2	167.7	391.7	340.9	307.6	203.4	85.1
1956	204.4	160.9	180.1	426.2	350.0	307.6	183.2	100.6
1957	204.4	183.1	173.9	426.2	377.6	340.8	203.4	112.1
1958	206.8	177.6	180.1	426.2	405.3	324.2	210.2	119.9
1959	228.4	188.8	204.9	437.7	442.2	365.8	257.8	135.4
1960	240.5	216.5	223.6	460.8	469.7	432.3	298.4	154.7
1961	262.1	233.2	236.0	495.3	488.3	448.9	312.0	166.3
1962	276.5	244.2	260.8	518.3	525.0	482.1	318.8	170.2
1963	295.8	277.6	291.9	564.4	580.4	490.4	345.8	185.6
1964	317.4	294.1	329.2	668.1	626.4	523.7	366.2	208.8
1965	317.4	310.8	378.9	691.1	681.7	548.6	407.0	224.3
1966	336.6	333.0	397.5	783.2	690.8	590.2	434.0	259.1
1967	363.1	360.7	410.0	840.8	727.8	623.4	454.4	282.2
1968	387.1	416.3	478.2	979.1	801.5	689.9	508.6	324.8

Year								
1969	442.4	505.1	565.2	1,036.7	875.0	806.3	583.2	359.6
1970	502.6	555.0	621.1	1,151.8	921.1	831.3	678.2	386.6
1971	536.3	571.7	683.2	1,220.9	967.3	798.0	732.4	409.9
1972	577.1	638.3	751.6	1,347.6	1,041.0	914.4	834.2	444.7
1973	556.3	686.8	844.0	1,489.1	1,108.7	979.3	922.4	508.3
1974	540.7	771.2	851.6	1,434.0	1,186.3	979.3	1,010.5	566.2
1975	612.1	728.9	793.7	1,335.1	1,137.6	808.9	969.8	501.1
1976	661.7	847.7	904.9	1,498.0	1,185.4	946.4	1,066.8	597.3
1977	659.1	874.8	949.2	1,631.3	1,234.0	1,043.9	1,125.0	621.2
1978	663.7	935.2	986.0	1,797.7	1,310.6	1,117.0	1,192.9	641.1
1979	724.7	1,025.9	1,040.2	1,824.7	1,444.3	1,222.0	1,312.2	671.9
1980	727.0	1,074.1	1,060.0	1,836.0	1,558.4	1,335.6	1,371.2	683.3
1981	700.8	1,112.9	1,062.2	1,902.1	1,592.7	1,375.7	1,408.3	728.4
1982	756.2	1,143.5	1,080.3	1,886.9	1,638.9	1,335.8	1,367.5	752.4
1983	744.8	1,195.0	1,124.5	2,026.5	1,715.9	1,389.2	1,414.0	716.3
1984	878.1	1,308.5	1,179.7	2,403.4	1,832.6	1,522.6	1,500.2	818.8
1985	958.9	1,436.7	1,226.8	2,595.7	1,964.6	1,536.3	1,554.2	867.1
1986	1,000.1	1,445.3	1,327.4	2,694.3	1,996.0	1,564.0	1,564.4	877.5
1987	1,094.1	1,484.3	1,419.0	2,882.9	2,057.9	1,589.0	1,617.7	902.9
1988	1,096.3	1,604.6	1,539.6	3,171.2	2,179.3	1,639.9	1,745.4	963.4
1989	1,128.1	1,779.4	1,670.5	3,139.5	2,312.2	1,677.6	1,900.7	1,060.8

Note: Figures refer to the customs territory of the year specified.

TABLE F.4. (Cont.) *Volume of Exports, Annual Data, 1950–1989 (1913 = 100)*

	Italy	Japan	Netherlands	Norway	Sweden	Switzerland	UK	USA
1950	126.5	210.1	171.2	269.5	275.9	113.2	100.0	224.6
1951	139.1	252.1	194.0	301.8	286.2	139.0	98.7	285.9
1952	126.5	252.1	205.5	280.3	255.5	133.1	92.0	299.5
1953	139.1	294.1	239.7	280.3	286.2	147.0	94.7	306.3
1954	151.8	336.2	273.9	323.4	306.6	154.9	98.7	299.5
1955	177.1	462.2	296.8	334.2	327.0	170.8	100.7	299.5
1956	202.4	546.3	308.1	377.3	357.7	188.7	112.0	353.9
1957	240.4	630.3	319.6	377.3	398.5	202.6	114.7	381.1
1958	253.0	672.3	342.4	377.3	388.4	198.6	110.7	326.7
1959	303.6	798.3	388.0	431.2	418.9	222.4	114.7	326.7
1960	366.9	924.4	445.1	463.5	480.3	245.8	120.0	387.9
1961	442.7	1,008.5	456.5	495.9	500.7	257.0	124.0	387.9
1962	493.4	1,176.6	490.8	528.2	541.5	273.7	126.7	408.4
1963	518.6	1,302.6	525.1	582.1	592.6	290.5	133.3	435.6
1964	594.5	1,638.8	593.5	679.1	654.0	312.8	137.3	496.8
1965	733.7	2,101.0	650.5	711.5	684.7	346.4	144.0	496.8
1966	834.9	2,395.1	684.8	776.2	735.8	374.3	149.3	530.9
1967	898.1	2,479.2	741.9	819.3	766.3	407.8	148.0	551.3
1968	1,062.6	3,109.5	867.5	927.1	838.0	463.7	168.0	592.1

Year								
1969	1,189.1	3,655.7	1,004.4	1,024.1	929.0	525.1	186.7	558.1
1970	1,265.0	4,202.0	1,141.3	1,078.0	1,021.8	558.6	192.0	680.6
1971	1,353.5	5,042.4	1,278.2	1,088.8	1,052.5	575.4	209.3	673.8
1972	1,543.3	5,378.6	1,403.8	1,239.7	1,113.7	608.9	213.1	735.1
1973	1,574.2	5,647.5	1,652.3	1,368.6	1,289.7	679.5	240.6	909.3
1974	1,689.1	6,613.3	1,701.8	1,357.7	1,340.0	677.5	253.3	988.4
1975	1,727.9	6,672.8	1,637.2	1,401.1	1,184.5	624.0	247.8	967.7
1976	1,938.7	8,140.8	1,850.0	1,628.1	1,222.4	699.5	269.1	1,001.5
1977	2,093.8	8,865.4	1,800.1	1,574.4	1,205.3	779.9	291.7	1,006.5
1978	2,328.3	8,962.9	1,861.3	1,941.2	1,322.2	817.3	299.6	1,116.3
1979	2,516.9	8,873.3	2,025.0	2,049.9	1,430.7	835.3	310.9	1,253.6
1980	2,303.0	10,355.1	2,045.3	2,168.9	1,402.0	849.5	313.4	1,337.6
1981	2,402.0	11,453.0	2,045.3	2,125.6	1,413.3	888.6	309.7	1,296.1
1982	2,411.6	11,190.0	2,024.8	2,104.4	1,459.9	854.8	318.4	1,158.7
1983	2,496.0	12,164.0	2,148.4	2,387.0	1,654.1	858.3	320.3	1,085.7
1984	2,660.8	14,110.2	2,293.0	2,590.0	1,779.8	961.3	353.3	1,159.5
1985	2,857.7	14,731.1	2,396.2	2,698.8	1,849.2	1,032.4	373.8	1,138.7
1986	2,909.1	14,657.4	2,446.5	2,763.6	1,904.7	1,043.8	386.6	1,205.9
1987	2,996.4	14,716.0	2,541.9	3,139.5	1,960.0	1,058.4	407.8	1,408.5
1988	3,167.2	15,348.8	2,750.4	3,343.5	1,024.6	1,132.5	423.7	1,697.2
1989	3,363.5	16,147.0	2,893.4	3,858.4	2,105.6	1,234.4	446.1	1,894.1

Note: Figures refer to the customs territory of the year specified.

TABLE F.5. *Value of Merchandise Exports f.o.b. at Current Prices and Exchange Rates* ($ million)

	1870	1913	1950	1973	1987
Australia	88	382	1,668	9,559	26,516
Austria	160	561	326	5,283	27,168
Belgium	133	717	1,652	22,455	83,098
Canada	58	421	3,020	26,437	98,168
Denmark	(40)	171	665	6,248	25,675
Finland	9	78	390	3,836	20,037
France	541	1,328	3,082	36,635	148,382
Germany	424	2,454	1,993	67,563	294,364
Italy	208	485	1,206	22,223	116,363
Japan	15	315	825	37,017	231,286
Netherlands	51	413	1,413	23,496	92,854
Norway	22	105	390	4,725	21,490
Sweden	41	219	1,103	12,201	44,506
Switzerland	(61)	266	894	9,528	45,515
UK	971	2,555	6,325	29,637	131,258
USA	403	2,380	10,282	71,404	254,484
Total	3,225	12,850	35,234	388,247	1,661,164

Source: Derived from sources cited in A. Maddison, *The World Economy in the Twentieth Century*, OECD, Paris, 1989.

TABLE F.6. *Merchandise Exports at 1985 Prices and Exchange Rates* ($ million)

	1870	1913	1950	1973	1987
Australia	318	2,372	3,764	13,193	25,947
Austria	580	2,512	800	8,241	17,742
Belgium	740	4,380	4,897	36,971	61,694
Canada	627	3,504	10,897	52,177	100,734
Denmark	183	870	2,083	9,645	17,745
Finland	172	886	1,768	8,681	14,390
France	2,035	6,542	9,761	60,342	104,095
Germany	3,755	21,213	7,382	107,823	190,425
Italy	1,039	2,685	3,396	42,261	80,286
Japan	36	1,203	2,527	67,922	176,989
Netherlands	1,087	2,849	4,877	47,067	72,269
Norway	193	741	1,996	10,134	22,183
Sweden	440	1,647	4,545	21,245	32,473
Switzerland	513	2,659	3,010	18,067	28,142
UK	8,424	27,087	27,087	65,172	110,245
USA	2,498	19,216	43,160	174,733	265,318
Total	22,640	100,364	131,950	743,674	1,319,677

Sources: Derived from volume indices presented above, and 1985 exports from IMF, *International Financial Statistics*.

TABLE F.7. *Ratio of Merchandise Exports to GDP at Current Market Prices*

	1913	1950	1973	1987
Australia	18.3	22.0	13.7	13.5
Austria	8.2	12.6	19.0	23.2
Belgium	50.9	20.3	49.9	59.8
Canada	15.1	17.5	20.9	23.9
Denmark	26.9	21.3	21.9	25.4
Finland	25.2	16.6	20.5	22.5
France	13.9	10.6	14.4	16.8
Germany	17.5	8.5	19.7	26.4
Italy	12.0	7.0	13.4	15.4
Japan	12.3	4.7	8.9	9.7
Netherlands	38.2	26.9	37.3	43.6
Norway	22.7	18.2	24.4	25.7
Sweden	20.8	17.8	23.5	27.6
Switzerland	31.4	20.0	23.2	26.6
UK	20.9	14.4	16.4	19.3
USA	6.1	3.6	8.0	5.7
Arithmetic average	21.2	15.1	20.9	24.1

Sources: A. Maddison, *The World Economy in the Twentieth Century*, OECD, Paris, 1989 for 1913–73. 1987 from IMF, *International Financial Statistics*, and OECD, *National Accounts*.

Switzerland: 1870–1900 from P. Bairoch, *Commerce extérieur et développement économique de l'Europe au XIXe siècle*, Mouton, Paris, 1976; 1939 onwards from UN, *Yearbook of International Trade Statistics*. Between 1959 and 1960 there is a break in the series and it was assumed that Swiss export prices rose 1 per cent (as in Germany) that year.

UK: 1700–1800 derived from P. Deane and W. A. Cole, *British Economic Growth 1688–1959*, Cambridge, 1964, pp. 319–21; 1800–1913 from A. H. Imlah, *Economic Elements in the Pax Britannica*, Harvard, 1958, pp. 94–8; 1913–70 from *The British Economy: Key Statistics 1900–1970*, London and Cambridge Economic Service, London, p. 14.

USA: 1790–1860 from D. C. North, *The Economic Growth of the United States 1790–1860*, Prentice-Hall, Englewood Cliffs, NJ, 1961, pp. 221

TABLE F.8. *Ratio of Merchandise Exports to GDP at 1985 Prices*

	1913	1950	1973	1987
Australia	10.9	7.7	9.5	12.4
Austria	5.2	4.0	12.6	20.0
Belgium	17.5	13.4	40.3	52.5
Canada	12.9	13.0	19.9	23.8
Denmark	10.1	9.3	18.2	25.8
Finland	17.0	12.7	20.5	23.0
France	6.0	5.6	11.2	14.3
Germany	12.2	4.4	17.2	23.7
Italy	3.3	2.4	8.7	11.2
Japan	2.1	2.0	6.8	10.6
Netherlands	14.5	10.2	34.1	40.9
Norway	14.6	13.5	27.4	34.0
Sweden	12.0	12.2	23.1	27.0
Switzerland	22.3	9.8	21.3	28.9
UK	14.7	9.5	11.5	15.3
USA	4.1	3.3	5.8	6.3
Arithmetic average	11.2	8.3	18.0	23.1

Source: Derived by dividing Table F.6 by GDP at 1985 US prices in Appendix A, Table A.2, or, where relevant, A.3.

and 241; 1860–78 estimated by converting the dollar value of exports into gold (the dollar was floating against gold until 1879) and then deflating by UK import price index. Gold price from G. F. Warren and F. A. Pearson, *Gold and Prices*, John Wiley, New York, 1935, p. 154; exports of US merchandise from *Historical Statistics of the U.S., Colonial Times to 1957*, US Dept. of Commerce, p. 538, adjusted to a calendar year basis; UK import price index back to 1870 from A. Maddison, 'Growth and Fluctuation', and 1860–70 from A. H. Imlah, *Economic Elements in the Pax Britannica*, Harvard, 1958, pp. 96–8. 1879–1950 from R. E. Lipsey, *Price and Quantity Trends in the Foreign Trade of the United States*, NBER, Princeton, 1963, Table A2.

INDEX OF SUBJECTS

agriculture 30, 31, 35, 37, 46, 57, 73, 74, 110, 125, 149–51, 175
asset lives 279–82

backwardness, relative 71
best practice 15, 23, 45.
BIS 168
Bretton Woods 132, 177, 179, 187, 190; *see also* international monetary system
business cycles 85–9

capital: capital stock 1, 11, 12, 13, 20, 22, 33, 38, 39, 47, 65, 66, 67, 70, 131, 139–46, 161, 278–92; embodiment of technical processes 22, 141–4; gross and net 140, 141, 142, 278–9; replacement, widening, deepening 23; residential 69, 70, 144–5; vintages 23, 142, 143, 165; *see also* human capital
capital output ratios 67, 70, 83
capitalism: capitalist development 5, 16–21, 48, 85, 110; 'advanced' 1, 8, 24, 167; breakdown (Marxian) 19; (Schumpeterian) 21, 104; 'late' 108, 127
catch-up 131
causality, ultimate: proximate 10, 11, 12, 133
Chinese stagnation 8, 10, 15, 16, 46
Club of Rome 18, 58
colonialism 26, 80–1, 83–4
conflict, social 17, 19
convergence, convergency process, divergence 48, 52, 130, 132, 160–4
currency 35, 39, 173, 178, 179; *see also* exchange rates, international payments system
cycle, *see* business cycles

demand management 168, 169–73, 180
demography, *see* fertility; migration; mortality; population
depressions 89, 97, 106, 111, 112
developing countries 26
divergence, *see* convergence

economic policy 34, 110, 167–92
economic structure 32, 36, 73, 74, 109, 110
economies of scale and specialization 15, 18, 23, 26, 42, 153
education 1, 23, 37, 63–5, 78, 82, 83, 131, 133, 138
embodiment of technical progress, *see* capital
employment, full 131, 169, 172, 173, 182, 185
energy 43, 59, 153–5, 185; *see also* oil, OPEC
entrepreneurs and entrepreneurship 20–2, 35, 103, 132
exchange controls 76, 179
exchange rates 177–80, 189; fixed 132, 173, 180, 189; floating 189, 190
exports, *see* trade

factor inputs 136–45; augmented 134, 135, 139, 146, 157, 158
fertility 13, 60, 241
fiscal policy 182–5
fluctuations, *see* business cycles; long waves
follower countries, *see* lead country

GDP, level and growth 22, 50, 57, 87–91, 95, 97, 105, 111, 112, 113, 115–19, 128, 129, 153, 155, 167, 180, 195–222
GDP per capita 6, 7, 8, 12, 24–5, 48, 49, 52, 54, 55, 57, 128, 129, 190, 191, 222
gold 176, 179, 180
gold standard 173
'golden age' 121, 123, 167, 192
government 76–80, 81, 103, 111, 169, 172, 173
growth accounts 11, 23, 26, 131, 132–66

human capital 12, 13, 23, 63–5, 82, 131; *see also* education

indicators of economic performance
    86, 87, 88, 94, 96, 97–108, 114–23,
    167
industrialization: 'industrial
    revolution' 110
inflation 168, 173, 185, 187, 188
innovations, *see* technical progress
institutions 10, 16, 33, 36, 52, 55–6, 69,
    110, 122, 131, 177
international monetary system
    111, 168–9, 177–9, 187, 189, 190
invention, *see* technical progress
investment 23, 33, 37, 39, 40, 41, 45,
    131, 173, 278; foreign 34, 39, 45; *see
    also* capital

labour hoarding 136–7
labour input, *see* labour supply
labour productivity, *see* productivity
labour supply 63, 134–9, 243–73
land 1, 14, 17, 40, 43, 58
lead country: leadership 1, 19, 21, 30,
    33, 34, 37, 40, 41, 43, 44, 45, 103,
    131; dichotomy leader–follower
    countries 23; follower countries 23,
    30, 44, 131, 140; leadership 23;
    British leadership (origins) 35–9;
    (loss) 39–40; Dutch leadership
    (origins) 30–4; (loss) 34–5; US
    leadership (origins and nature)
    40–4, 65, 69, 72, 131; (waning) 44
long swings, *see* long waves
long waves 85, 89–111

Marshall Plan 123, 173; Marshall
    Aid 152
mercantilism, mercantilist policy 34,
    36, 76
migration 240
mortality 60

natural resources 16, 20, 40, 44, 56–8,
    60, 82, 155–9
NBER 86, 87, 101, 102, 114

oil 58, 132, 155, 166, 177, 181–2, 185
OPEC 105, 132, 147, 155, 181, 182,
    185, 190

phases of growth 111–24
policy, *see* economic policy

pollution 58
population 9, 10, 12, 13, 26, 31, 34, 57,
    60–3, 78, 82, 96, 101, 128, 129,
    223–45; *see also* fertility, migration,
    mortality
prices 88, 94–6, 111, 119, 122, 126,
    154, 167, 170–2, 174–7, 179–81,
    293–307; *see also* inflation
productivity 71, 128, 131, 132, 149;
    joint factor 37, 38, 71; labour (GDP
    per man hour) 30, 31, 38, 39, 43, 45,
    46, 47, 51, 53, 71, 129, 130, 258, 259,
    274–7; sectoral 150–1
protocapitalism, protocapitalist 8, 11–
    16, 18
purchasing power parity (PPP) 7, 25,
    66, 196–7, 199, 281, 285–7, 289, 290

recessions, *see* business cycles
research and development (R and
    D) 1, 21, 42, 43, 153
reserves, *see* international monetary
    system

slow-down 128, 129, 131, 132, 159–61
socialism, 'socialist' 19, 80
stagnation 18, 34
structure and structural change 73–4,
    132, 133, 148–52
system shocks 108, 110, 177

'take-off' 82, 105
technical progress 11–15, 17–21, 23,
    33, 36, 37, 66, 69, 72, 141–4, 146,
    152, 165; technological diffusion
    152–3
trade: foreign and international 15, 17,
    34, 35, 46, 67, 74–6, 88, 92–3, 131,
    132, 133, 147–8, 169, 180, 308–27

unemployment 19, 119, 122, 123, 136,
    138, 169, 170–1, 176, 192, 194,
    253–5, 260–5

vintages, *see* capital

working hours 19, 63, 82, 137, 255–8,
    270–3

Zollverein 74

INDEX OF AUTHORS

Abel, W. 27, 28
Abramovitz, M. 96, 101, 126
Aukrust, O. 194, 220, 255

Baba, M. 310
Bairoch P. 32, 46, 202, 227, 245, 249,
  251, 308, 326
Balke, N. S. 221
Baudhuin, F. 293
Baumol, W. J. 83
Bell, D. 176, 193
Beloch, E. J. 224
Beveridge, W. H. 193
Bienefeld, M. A. 82, 255
Bjerke, K. 293, 309
Björkqvist, M. 293
Blackman, A. B. 83
Blades, D. W. 83, 201, 205
Blanchard, O. 184, 194
Blaug, M. 28
Bochove, A. van 205, 288
Boserup, E. 28
Bresciani-Turroni, C. 310
Bronfenbrenner, M. 125
Brown, A. J. 193
Bry, G. 294
Buckley, J. E. 258
Burns, A. F. 72, 125, 126
Butlin, M. W. 201, 308
Butlin, N. G. 201, 251, 293, 308

Calmfors, L. 194
Capelle, S. 308
Carbonnelle, C. 202
Chambers, J. D. 45
Chandler, A. D. Jnr. 47
Cipolla, C. M. 27, 45
Clark, C. 22, 58, 220
Crafts, N. F. R. 27, 220
Crosland, C. A. R. 176, 193
Crouzet, F. 203, 309

David, P. A. 220
Deane, P. 46, 220, 225, 326
Denison, E. F. 23, 26, 29, 64, 83, 153,
  165, 204, 258
Dessirier, J. 203

Dewhurst, J. F. 59, 227
Dickson, D. 225
Dobb, M. 28
Dornbusch, R. 184, 194
Dosi, G. 166
Duin, J. J. van 126

Easterlin, R. A. 82, 126
Edgren, G. 194
Eichengreen, B. 254
Elvin, M. 27, 46
Emminger, O. 179, 193
Evans, A. A. 258

Faber, J. A. 224
Faxen, K. O. 194
Feinstein, C. H. 32, 41, 82, 220, 224,
  252, 254, 255, 277, 203, 288, 292
Feis, H. 46
Fenoaltea, S. 204
Firestone, O. J. 202, 240
Fogel, R. W. 165
Forrester, J. W. 126
Flora, P. 77, 83
Fontvieille, L. 77
Freeman, C. 152, 166
Friedman, M. 177, 182, 193
Fua, G. 204, 294, 310

Gadisseur, J. 202
Galbraith, J. K. 176, 193, 204
Galenson, W. 254
Gallman, R. E. 83, 221
Garry, G. 126
Gelderen, J. van (under pseudonym
  J. Fedder) 106, 126
Gerschenkron, A. 82
Giavazzi, F. 194
Gilbert, M. 22
Gille, H. 61, 227
Glass, D. V. 241
Goldsmith, R. W. 41
Goldstein, J. S. 126
Golinelli, R. 204
Good, D. F. 293
Goossens, M. 202
Gordon, R. J. 221

Granier, R. 252, 255
Griliches, Z. 166

Habakkuk, H. J. 241
Hansen, B. 193
Hanley, S. B. 27, 205, 221, 222
Hansen, S. A. 202, 223, 251, 252, 254, 309
Hartog, H. den 288
Hathaway, D. E. 193
Hatton, T. J. 254
Hayami, A. 224
Hayami, Y. 47, 82
Hayek, F. A. von 182
Hedges, J. N. 258
Henry, L. 28, 223
Heston, 25, 201
Hjerppe, R. 203
Hobsbawm, E. J. 82
Hoffmann, W. G. 140, 203, 224, 240, 294, 309

Imlah, A. H. 326, 327

Jevons, W. S. 58, 82
Jewkes, J. 83
Johansson, O. 310
Jones, E. L. 27
Jörberg, L. 294
Juglar, C. 85, 86, 102, 125

Kaldor, N. 83
Kausel, A. 202, 223, 251, 254
Keating, M. 251, 254
Kendrick, J. W. 32, 41, 221, 256
Keynes, J. M., and Keynesianism 172, 182, 184
Kindleberger, C. P. 308
King, G. 14, 28, 45
Kirner, W. 41, 288
Kitchin, J. 102, 125, 126
Kneschaurek, F. 220
Kojima, K. 310
Kondratieff, N. D. 2, 89, 94, 95, 96, 102, 103, 104, 106, 111, 125, 126
Kosai, Y. 205
Krantz, O. 220
Kraus, F. 83
Krause, L. B. 193
Kravis, I. B. 22
Kuznets, S. 10, 22, 27, 41, 72, 83, 89, 96, 97, 101, 102, 111, 118, 122, 125, 126, 127, 240

Lal, D. 27
Landes, D. S. 27, 45, 46
Laslett, P. 28
Layard, R. 184, 194
Lebergott, S. 252, 254
Lee, J. 225
Leroy Ladurie, E. 27, 28
Levasseur, E. 309
Levy-Leboyer, M. 41, 203, 309
Lewis, W. A. 46, 96, 97, 125
Lindblad, J. T. 310
Lindert, P. H. 32
Lipsey, R. G. 327
List, F. 19
Lützel, H. 282

McCloskey, D. N. 46, 47
McCracken, P. M. 193
McCusker, J. J. 46
Malinvaud, E. 41, 140, 251, 255, 284
Malthus, T. R., and Malthusianism 1, 11, 12, 13, 16, 27, 58
Mandel, E. 105–8, 127
Marczewski, J. 293, 294
Marx, K. 1, 17–19, 21, 28
Mathias, P. 203
Matthews, R. C. D. 82
Mayer, K. B. 224, 241
Meadows, D. H. 82
Mendelbaum, S. 256
Mensch, G. 127
Menshikov, S. M. 127
Mill, J. S. 18
Mingay, G. E. 46
Mitchell, B. R. 46, 225, 240, 241, 293, 294, 308, 310, 311
Mitchell, W. C. 125, 126
Mueller, B. 241
Mulhall, M. G. 311
Musson, A. F. 46
Myrdal, G. 176, 193, 294

Needham, J. 8, 27
Nelson, R. R. 47
North, D. C. 326
Nolz, E. 294
Nurkse, R. 193

Odling-Smee, J. C. 82
Ohdner, C. E. 194
Ohkawa, K. 41, 70, 140, 148, 205, 288, 310
Olgaard, A. 309

Olson, M. 27
Oomens, C. A. 224
Oppenheimer, P. M. 193

Perkins, D. H. 8, 10, 27
Phelps, E. S. 193
Phelps Brown, F. H. 47
Pihkala, E. 309
Pilat, D. 47
Postan, M. M. 203, 241
Postlethwaite, T. N. 83

Rapp, R. 28, 45
Reijnders, J. P. G. 127
Ricardo, D. 1, 17, 18, 28, 58
Riley, J. G. 45
Ringstad, V. 166
Robinson, E. 46
Rolfe, S. E. 193
Romer, P. M. 153, 166
Rosovsky, H. 41, 148, 288, 310
Rostow, W. W. 27, 82, 105, 127
Ruttan, V. W. 47, 82

Salant, W. S. 194
Salter, W. E. G. 23, 29, 165
Sauvy, A. 203
Schmidt, L. 282
Schröder, W. H. 82, 127
Schultz, T. W. 23, 29, 82
Schumpeter, J. A. 1, 2, 17, 20, 28, 29,
    82, 83, 89, 102–5, 106, 111, 126
Senior, N. W. 63
Shinohara, M. 70, 140, 205, 288
Slicher van Bath, B. H. 32
Smith, A. 11, 15, 16, 21
Smith, T. C. 205
Solomou, S. 127
Solow, R. L. 28
Solow, R. M. 29, 125, 165
Sombart, W. 103
Sorrentino, C. 254
Spiethoff, A. 82, 102

Spree, R. 127
Strachey, J. 127
Summers, R. 25, 201
Svennilson, I. 224

Taeuber, J. B. 224
Thomas, B. 126
Thorp, W. L. 125
Tilly, R. H. 203
Timmer, C. P. 46
Toutain, J.-C. 77, 203
Tuganl-Baranowsky, M. von 82, 86,
    102, 125

Umemura, M. 251, 252
Urlanis, B. T. 10, 227
Urquhart, M. C. 202, 223, 241, 251,
    293, 309

Vitali, O. 251
Volcker, P. 184
Vries, Jan de 27, 32, 45, 46, 205, 227
Vries, Johan de 75

Warren, G. F. 327
Weber, M. 103
Wee, H. van der 45
Williamson, J. 193
Williamson, J. G. 32, 126
Wolf, J. 83
Wolff, F. N. 83
Woodruff, W. 240
Woude, A. M. van der 32
Wright, G. 47
Wrigley, E. A. 225

Yamamura, K. 205

Zanden, J. L. van 45, 205, 294, 310
Zeeuw, J. W. de 45
Zellner, A. 254
Zwingli, U. 220